AF891952

Miriam Schönbein

Omnidirectional Stereo Vision for Autonomous Vehicles

Schriftenreihe
Institut für Mess- und Regelungstechnik,
Karlsruher Institut für Technologie (KIT)

Band 032

Eine Übersicht aller bisher in dieser Schriftenreihe erschienenen
Bände finden Sie am Ende des Buchs.

Omnidirectional Stereo Vision for Autonomous Vehicles

by
Miriam Schönbein

KIT Scientific Publishing

Dissertation, Karlsruher Institut für Technologie (KIT)
Fakultät für Maschinenbau, 2014
Tag der mündlichen Prüfung: 19. Dezember 2014
Referenten: Prof. Dr.-Ing. Christoph Stiller
Prof. Dr.-Ing. Jürgen Beyerer

Impressum

KIT Scientific Publishing

Karlsruher Institut für Technologie (KIT)
KIT Scientific Publishing
Straße am Forum 2
D-76131 Karlsruhe

KIT Scientific Publishing is a registered trademark of Karlsruhe Institute of Technology. Reprint using the book cover is not allowed.

www.ksp.kit.edu

This document – excluding the cover – is licensed under the Creative Commons Attribution-Share Alike 3.0 DE License (CC BY-SA 3.0 DE): http://creativecommons.org/licenses/by-sa/3.0/de/

The cover page is licensed under the Creative Commons Attribution-No Derivatives 3.0 DE License (CC BY-ND 3.0 DE): http://creativecommons.org/licenses/by-nd/3.0/de/

Print on Demand 2015

ISSN 1613-4214
ISBN 978-3-7315-0357-6
DOI: 10.5445/KSP/1000046298

Omnidirectional Stereo Vision for Autonomous Vehicles

Zur Erlangung des akademischen Grades

Doktor der Ingenieurwissenschaften

der Fakultät für Maschinenbau
des Karlsruher Instituts für Technologie (KIT)
genehmigte

Dissertation

von

DIPL.-ING. MIRIAM SCHÖNBEIN

aus Karlsruhe

Tag der mündlichen Prüfung: 19. Dezember 2014
Hauptreferent: Prof. Dr.-Ing. Christoph Stiller
Korreferent: Prof. Dr.-Ing. Jürgen Beyerer

Vorwort

Die vorliegende Arbeit entstand während meiner Tätigkeit als wissenschaftliche Mitarbeiterin am Institut für Mess- und Regelungstechnik am Karlsruher Institut für Technologie (KIT). Zunächst möchte ich mich herzlich bei Herrn Prof. Dr.-Ing. Christoph Stiller für die Betreuung dieser Arbeit, die Freiheiten bei der Ausgestaltung des Themas und die vielen fachlichen Diskussionen bedanken. Herrn Prof. Dr.-Ing. habil. Jürgen Beyerer danke ich für die Übernahme des Korreferats und das damit verbundene Interesse an meiner Arbeit.

Bei meinen Kollegen und Kolleginnen am Institut möchte ich mich für die gute Zusammenarbeit in sehr angenehmer Arbeitsatmosphäre, für die zahlreichen Diskussionen in der Kaffeerunde und bei den Sommerseminaren und für die gemeinsamen Freizeitaktivitäten wie Grillen und Skifahren bedanken. Besonders bedanken möchte ich mich bei meinen Scrumpartnern Holger Rapp, Henning Lategahn und Philip Lenz. Ein besonderer Dank gilt auch Andreas Geiger für die Unterstützung und Zusammenarbeit auch über die gemeinsame Zeit am MRT hinaus. Martin Lauer danke ich für die Einführung in das Themenfeld der katadioptrischen Kameras. Ebenso bedanken möchte ich mich bei Eike Rehder, Martin Lauer, Johannes Beck, Henning Lategahn, Philip Lenz und Markus Schreiber für die mühevolle Arbeit des Korrekturlesens und für die konstruktiven Hinweise zu meiner Arbeit.

Des Weiteren möchte ich mich bei unserem Sekretariat für die jederzeit gute Hilfe bei Verwaltungsaufgaben und die weibliche Unterstützung bedanken. Ein weiterer Dank gilt den Werkstätten und unserem Systemadministrator für die zuverlässige Hilfe in allen praktischen Belangen. Für die finanzielle Unterstützung und die fachlichen Diskussionen auch aus anderen Bereichen danke ich der Karlsruher School of Optics and Photonics (KSOP).

Mein ganz besonderer Dank gilt meinen Eltern Friederike und Rainer und meinem Freund Markus, die mich sowohl in fachlichen als auch nicht fachlichen Belangen immer unterstützt haben und mir in schwierigen Zeiten zur Seite standen.

Karlsruhe, im Dezember 2014 *Miriam Schönbein*

Abstract

Environment perception is an important requirement for many applications for autonomous vehicles and robots. Cameras are often used to provide visual information similar to the human vision system. However, conventional perspective cameras typically used for environment perception have only a very limited field of view.

In this thesis, we present a stereoscopic omnidirectional camera system for autonomous vehicles which resolves the problem of a limited field of view. The proposed setup consists of two horizontally aligned catadioptric cameras mounted on top of a vehicle and provides a 360° panoramic view of the environment. We show that this camera setup overcomes major drawbacks of traditional perspective cameras in many applications for autonomous systems.

However, due to misalignments between camera and mirror, catadioptric camera systems are slightly non-central systems even if they are designed to fulfill the single viewpoint (SVP) condition. There exist two types of projection models for catadioptric cameras: Central models which have very cheep computational time but assume that the camera systems fulfill the SVP condition, and non-central models which are very accurate but not efficient enough. We propose a novel projection model for slightly non-central cameras which is both, very accurate and efficient at the same time. Moreover, a calibration toolbox to calibrate stereoscopic catadioptric cameras with the proposed projection model was designed. In contrast to existing toolboxes, the developed calibration toolbox allows for calibrating multiple catadioptric cameras with different projection models. We show the benefits of the proposed projection model with extensive experiments evaluated regarding the calibration results compared to several other projection models.

Based on the proposed setup and projection model, we present an ego-motion estimation with catadioptric cameras which yields high precision estimates. Beyond that, a comparative study of feature matching strategies which is an input for the ego-motion estimation is given. The precise motion estimation is used to create high fidelity top view maps of the driven path and the nearby surrounding. Furthermore, we present an approach to obtain dense 360° panoramic depth images and a dense 3D reconstruction of the environment. The proposed approach uses the stereoscopic catadioptric setup only and combines motion and spatial stereo for dense 3D information.

Kurzfassung

Visuelle Umfeldwahrnehmung ist eine elementare Anforderung für viele Anwendungen im Bereich autonomer Fahrzeuge und Roboter. Zur visuellen Erfassung der Umgebung werden oft Kameras eingesetzt, die eine Wahrnehmung ähnlich derer des Menschen ermöglichen. Typischerweise verwendete, fest eingebaute perspektivische Kameras verfügen jedoch nur über einen sehr beschränkten Sichtbereich.

In dieser Arbeit wird ein stereoskopisch-omnidirektionales Kamerasystem für autonome Fahrzeuge vorgestellt, das das horizontale Sichtfeld nicht einschränkt. Das vorgestellte Kamerasystem besteht aus zwei oben auf dem Fahrzeug angebrachten und horizontal ausgerichteten katadioptrischen Kameras und ermöglicht eine 360° Rundumsicht. Dieses Kamerasystem wird im Rahmen der Arbeit auf seine Eignung für mehrere Anwendungen im Fahrzeug evaluiert und erzielt dabei klare Verbesserungen im Vergleich zu traditionellen perspektivischen Kameras.

Aufgrund von Verschiebungen zwischen Kamera und Spiegel verletzen katadioptrische Kamerasysteme die Annahme eines effektiven Projektionszentrums (“single viewpoint”, SVP) und werden daher als nicht-zentrale Systeme bezeichnet. Auch Systeme die dazu ausgelegt werden die SVP-Bedingung zu erfüllen, verletzen diese normalerweise zumindest geringfügig aufgrund von Fertigungstoleranzen. Es existieren zwei Arten von Projektionsmodellen für katadioptrische Kameras: Zentrale Modelle, die zwar nur geringe Rechenzeit benötigen aber voraussetzen, dass die Kameras die SVP-Bedingung erfüllen, und nicht-zentrale Modelle, die eine sehr genaue Abbildung erlauben aber entsprechend aufwendig zu berechnen sind. In dieser Arbeit wird ein neuartiges Projektionsmodell für geringfügig von der SVP-Bedingung abweichende Systeme vorgestellt, das sehr genau abbildet und gleichzeitig nur geringe Rechenzeit benötigt. Im Rahmen dieser Arbeit wurde darüber hinaus eine Toolbox zur Kalibrierung stereoskopisch-katadioptrischer Kameras mit dem vorgestellten Projektionsmodell entwickelt. Im Vergleich zu existierenden Ansätzen erlaubt die vorgestellte Toolbox die Kalibrierung mehrerer katadioptrischer Kameras zueinander und erlaubt die Anwendung verschiedener Projektionsmodelle. Die Vorteile des vorgestellten Projektionsmodells werden durch umfangreiche Experimente bestätigt, in dem die Kalibrierergebnisse bezüglich verschiedener Projektionsmodelle ausgewertet werden.

Basierend auf dem vorgeschlagenem Aufbau und dem Projektionsmodell wird ein Algorithmus zur Bewegungsschätzung mittels katadioptrischer Kameras vorgestellt, der hoch genaue Positionsschätzungen erlaubt. Ein Vergleich verschiedener Strategien zum Finden von Merkmalskorrespondenzen, die als Eingangswerte für die Bewegungsschätzung nötig sind, wird gezeigt. Anschließend wird die präzise Bewegungsschätzung zur Erstellung einer hoch genauen Draufsichtskarte der befahrenen Strecke genutzt. Weiterhin wird eine Methode vorgestellt um dichte 360° Rundumsicht-Tiefenbilder und die resultierende dichte 3D Rekonstruktion der Umgebung zu erhalten. Das vorgestellte Verfahren nutzt ausschließlich den stereoskopisch-katadioptrischen Aufbau und kombiniert zeitliches und räumliches Stereo für die dichte 3D Information.

Contents

Notation and Symbols

Acronym

2D/3D	**2/3-d**imensional
BM	**B**lock **M**atching
BRIEF	**B**inary **R**obust **I**ndependent **E**lementary **F**eatures
BRISK	**B**inary **R**obust **I**nvariant **S**calable **K**eypoints
CS	**C**oordinate **S**ystem
DOF	**D**egree **o**f **F**reedom
FAST	**F**eatures from **A**ccelerated **S**egment **T**est
IMU	**I**nertial **M**easurement **U**nit
FOV	**F**ield **o**f **V**iew
GPS	**G**lobal **P**ositioning **S**ystem
ORB	**O**riented FAST and **r**otated **B**RIEF
RANSAC	**Ran**dom **Sa**mpling **C**onsensus
SGM	**S**emi-**G**lobal **M**atching
SIFT	**S**cale **I**nvariant **F**eature **T**ransform
SLAM	**S**imultaneous **L**ocalization **a**nd **M**apping
SURF	**S**peeded **u**p **R**obust **F**eatures
SVP	**S**ingle **V**iew**p**oint
WTA	**W**inner **T**akes **A**ll

General Notation

Scalars	Regular (greek) lower case	a, b, c, σ, λ
Vectors	Bold (greek) lower case	$\mathbf{a}, \mathbf{b}, \mathbf{c}, \boldsymbol{\sigma}, \boldsymbol{\lambda}$
Matrices	Bold upper case	$\mathbf{A}, \mathbf{B}, \mathbf{C}$
Sets	Calligraphic upper case	$\mathcal{A}, \mathcal{B}, \mathcal{C}$

Projection

$\mathbf{F}$	Mirror focal point
$\mathbf{p} = [x, y, z]^\mathsf{T}$	3D world point in Cartesian coordinates
$\mathbf{p} = [\rho, \varphi, \theta]^\mathsf{T}$	3D world point in spherical coordinates
$\mathbf{q}^{(*)} = [u, v]^\mathsf{T}$	2D image point

where $(*) \in \{S, D, G, C, P, E\}$ denotes the projection model with S = Sphere Camera Model, D = Polynomial Distortion Model, G = Geometric Model, C = Centered Model, P = Perspective Model and E = estimated from the corner extraction.

$\mathbf{K}$	Intrinsic projection matrix
f_u, f_v	Focal lengths
c_u, c_v	Principal point
α	Skew parameter
$\mathbf{k} = [k_1, \ldots, k_5]^\mathsf{T}$	Distortion parameters

Sphere Camera Model

ξ, η	Sphere mirror parameters
$\zeta_i = f_i \eta$	Sphere focal lengths (with $i \in \{u, v\}$)
$\mathbf{p}_s$	3D point on the unit sphere
$\mathbf{p}_\xi$	3D point on the unit sphere in mirror coordinates
$\mathbf{q}_u^{(S)}$	Projected point into the normalized plane
$\mathbf{q}_d^{(S)}$	Projected undistorted point

Polynomial Distortion Model

$\mathbf{p}_p$	Vector through the world point $\mathbf{p}$
$\mathbf{q}_s^{(D)}$	2D point on the sensor plane
$f^{(D)}(\nu)$	Polynomial function with $\nu = \lVert\mathbf{q}_s^{(D)}\rVert$
$d_0, \ldots, d_n$	Polynomial distortion model parameters
$\mathbf{A}_S \in \mathbb{R}^{2\times 2}$	Affine transformation matrix
$\mathbf{t}_S \in \mathbb{R}^2$	Affine transformation vector

Non-Central Geometric Model

A, B, C	Spherical mirror parameters
a, b	Mirror parameters

$\mathbf{m} = [x_m, y_m, z_m]^\mathsf{T}$	Reflection point on the mirror surface
$\mathbf{c} = [x_c, y_c, z_c]^\mathsf{T}$	Camera location

If the points have no index they are represented in the mirror coordinate system. The indices C and R denote the camera coordinate system and the rotated mirror coordinate system, respectively.

$\mathbf{w}_r$	Mirror reflection ray
$\mathbf{w}_c$	Mirror incoming ray
$\mathbf{n}$	Normal vector on the mirror surface
$\mathbf{s}$	Intersection point between surface normal and z-axis
Π	Intersection plane
$\mathbf{n}_\Pi$	Normal vector of the intersection plane
$\mathbf{R}_R$	Pre-rotation matrix
$\mathbf{R}_C, \mathbf{t}_C = \mathbf{c}$	Transformation between mirror and camera coordinates
f	Polynomial equation
f_R	Polynomial equation with rotated parameters
$a_0, \ldots, a_8$	Polynomial coefficients
$\mathbf{q}_n^{(G)} = [x_n, y_n]^\mathsf{T}$	Normalized geometric projection
$\mathbf{q}_d^{(G)}$	Distorted normalized geometric projection

Centered Model

φ, θ	Viewing ray angles
$\mathbf{v}$	Optimal single viewpoint
$\mathbf{b} = (b_1, \ldots, b_k)$	Polynomial central-centered model coefficients
$Q(\varphi, \theta)$	Central-centered projection function

Calibration

$\mathbf{H} \in \mathbb{R}^{4\times 4}$	Transformation in homogeneous coordinates
$\mathbf{H}_{ex} = \begin{bmatrix} \mathbf{R}_{ex} & \mathbf{t}_{ex} \\ \mathbf{0} & 1 \end{bmatrix}$	Transformation between left and right camera (extrinsics)
$\mathbf{r}_{ex}$	Extrinsic rotation vector
$\mathbf{q}_{ex}$	Extrinsic quaternion vector
$\mathbf{H}_{cb} = \begin{bmatrix} \mathbf{R}_{cb} & \mathbf{t}_{cb} \\ \mathbf{0} & 1 \end{bmatrix}$	Transformation between checkerboard and mirror coordinate system
r_I	Mirror radius in the image
r_M	Mirror radius

Ego-motion

The indices l and r denote the left and right camera of the stereo rig, respectively. The index t denotes the current and $t-1$ the previous frame.

$\mathbf{H}_M$	Transformation between two frames in the reference camera (left)
$\mathbf{H}_{imu}$	Transformation between two frames in the IMU coordinate system
$\mathbf{H}_{cam,velo}$	Transformation between the left camera and the Velodyne
$\mathbf{H}_{velo,imu}$	Transformation between Velodyne and IMU
λ	Parameter (one or two cameras)
$e_i(j)$	End-point error for frame i starting at frame $i-j$

Top View Map

$\mathbf{p}_{vip}$	3D point on the virtual perspective image plane
u_p, v_p	Pixels on the virtual perspective image plane
φ_0, θ_0	Position for the virtual image plane
f_p	Virtual focal length

Reconstruction

$e_{11}, e_{12}, e_{21}, e_{22}$	Epipoles on the sphere
$\mathbf{E}_{12}$	Essentiell matrix between two frames
γ	Spherical disparity
$\mathbf{p}_S = [\rho_S, \varphi_S, \theta_S]^\mathsf{T}$	World point in the spherical rotated coordinate system
$\mathbf{R}_S$	Rotation matrix between rotated and original coordinate system

Index 1 denotes the first camera of the image pair and index 2 the second camera, respectively.

$\|\|\mathbf{t}\|\|$	Baseline between the cameras
$\mathbf{H}_V = \begin{bmatrix} \mathbf{R}_V & \mathbf{t}_V \\ \mathbf{0} & 1 \end{bmatrix}$	Transformation virtual coordinate system
$\mathbf{p}_V$	3D world point in the virtual coordinate system
$\mathbf{q}_V$	2D image point in the virtual coordinate system
$r = \sqrt{x_V^2 + y_V^2}$	Depth in the panoramic inverse depth image
$D = 1/r$	Inverse depth

Plane Estimation

d_h	Horizontal plane parameter (distance)
d_v, α_v	Vertical plane parameters (distance and angle)
E_u	Unary term
E_p	Pairwise term
$\mathcal{S} = \{s_1, \ldots, s_M\}$	Superpixel plane correspondence
$s \in \{1, \ldots, N\}$	Discrete plane index
M	Number of superpixels
N	Number of planes
$\mathcal{N}_S$	Set of neighboring superpixels
$\mathcal{Q}_s$	Set of valid inverse depth hypotheses
$\mathcal{H}$	Set of horizontal planes
w_{u_1}, w_{u_2}	Unary weight parameters
w_p	Smoothness parameter
τ_u, τ_p	Truncation parameters
D_{gt}	Velodyne ground truth inverse depth
D_{est}	Estimated inverse depth
e	Inverse depth error

Chapter 1

Introduction

Autonomous mobile robots are becoming more and more popular, particularly in the field of service and industrial robots. In the field of mobility, driver assistance systems, e.g., adaptive cruise control, lane departure warning, drowsiness detection, self-parking and many more, support the driver and help to reduce the number of accidents [122]. Consequently, a future step will be fully autonomous vehicles which will not only reduce the number of accidents but also prevent traffic jams and reduce pollution. There have already been many advances in autonomous driving such as the recently presented Bertha Benz Drive [132], the Google Driverless Car, and the approaches from the DARPA Urban Challenge [14]. First contributions are already presented by Dickmanns et al. [26] in the 1990s.

For autonomous systems in complex scenarios, environment perception is a very important task, e.g., for detecting objects or other traffic participants or for determining the own position in the world. Autonomous systems are usually equipped with multiple sensor types such as radar sensors, ultrasonic sensors, and particularly cameras to cover as much of the environment as possible. Thereby, cameras have the advantages of low cost and small construction space. Furthermore, they provide visual information similar to humans eyes. However, commonly used perspective cameras capture only a limited field of view. Moreover, these cameras typically point in frontal direction and objects alongside the sensor platform are not visible with a single perspective camera.

A panoramic view of the environment is desirable for autonomous systems since lateral objects often interfere with the ego-vehicle. There are important applications for autonomous driving and advanced driver assistance systems which also require side view such as blind spot detection or lane change assistance. In addition, a panoramic view of the surrounding can improve existing applications as localization, object detection, or lane tracking and enable new applications such as intersection reconstruction for autonomous driving in urban environments. Omnidirectional cameras similar to those used in this work are able to provide such information.

1.1. Omnidirectional Vision

Omnidirectional cameras overcome the problem of the limited field of view of standard perspective cameras and provide a panoramic image of the environment. There are several ways to obtain an omnidirectional image:

- From multiple images (mosaicing),
- from cameras with wide angle lenses (fisheye), or
- with the combination of a convex mirror and a lens (catadioptric).

The possibility to create a panoramic image by mosaicing a sequence of images can be accomplished from one rotating camera or from multiple fixed cameras. A common camera system for panoramic images obtained from multiple cameras is the *PointGrey Ladybug*. This camera system consists of six single lenses in one construction place as shown in Fig. 1.1c. Such systems are capable to obtain a high spatial resolution image as depicted in Fig. 1.1f. However, the possibility to obtain a panoramic view of the environment of the vehicle by capturing a set of perspective images suffers from the complexity of stitching the images together, the extensive cross-calibration of all cameras, and the required space to mount all cameras. Moreover, violations of the single viewpoint condition while stitching a panoramic image from multiple images introduces undesirable effects such as ghosting. Hence, these systems are not suitable for the dynamic environment of a vehicle.

One possibility to obtain an omnidirectional image with a single shot are special shaped wide angle lenses which capture a field of view of approximate 180° (e.g., fisheye lenses shown in Fig. 1.1b). However, these systems do not provide a complete panoramic image of the environment with a single camera (Fig. 1.1e).

The third possibility to obtain an omnidirectional image is the combination of a shaped mirror in front of a normal camera lens as shown in Fig. 1.1a. These systems provide a 360° field of view with one single shot (Fig. 1.1d). This work focuses on omnidirectional cameras composed of a

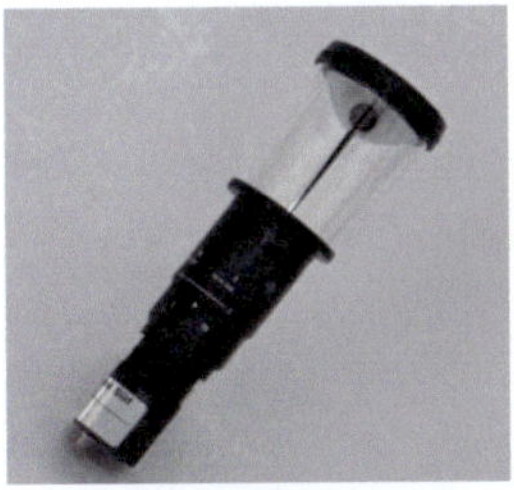

(a) Catadioptric

(b) Fisheye

(c) Ladybug

(d) Catadioptric Image

(e) Fisheye Image

(f) Ladybug Image

Figure 1.1.: **Omnidirectional Cameras.** This figure shows different camera systems to obtain an omnidirectional image of the surrounding and their captured images, particularly a catadioptric camera (a) providing a 360° field of view (FOV) (d), a fisheye camera (b) providing a 180° FOV (e), and a Ladybug camera (c) providing a stitched panoramic image (f).

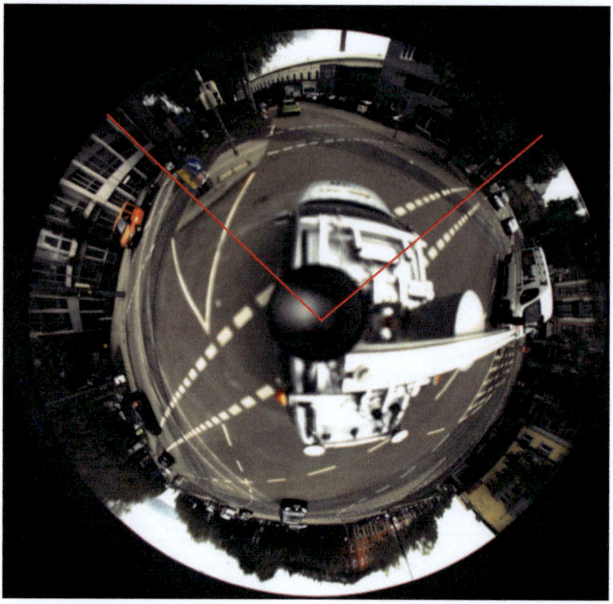

(a) Catadioptric Image (Resolution 1400×1400, FOV = 360°)

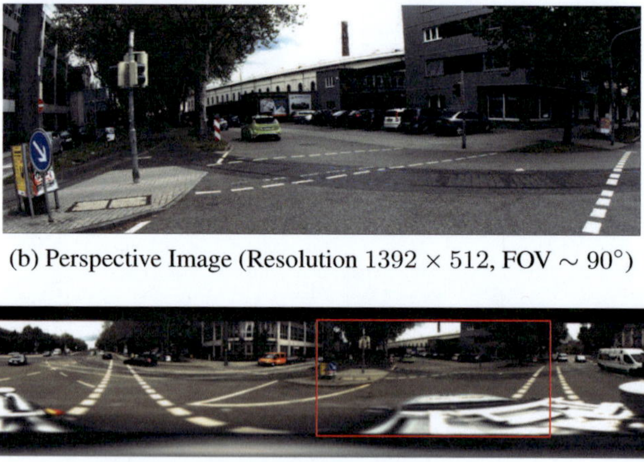

(b) Perspective Image (Resolution 1392×512, FOV $\sim$ 90°)

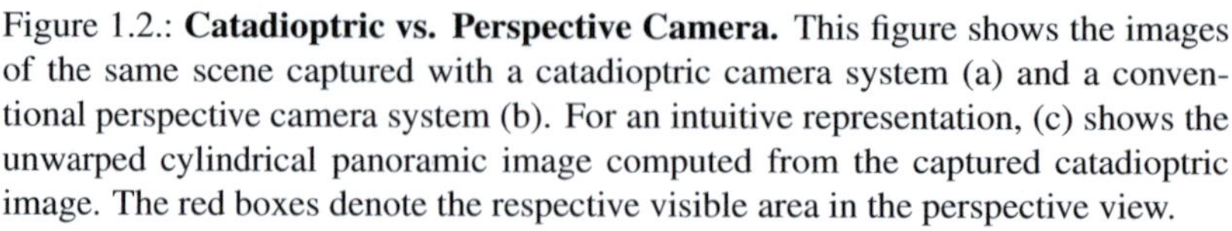

(c) Panoramic view of the same intersection scene

Figure 1.2.: **Catadioptric vs. Perspective Camera.** This figure shows the images of the same scene captured with a catadioptric camera system (a) and a conventional perspective camera system (b). For an intuitive representation, (c) shows the unwarped cylindrical panoramic image computed from the captured catadioptric image. The red boxes denote the respective visible area in the perspective view.

mirror and a lens. Such systems combine the principles of refraction (dioptric) and reflection (catoptric) in one single optical system and are called catadioptric cameras. The idea to use refractive as well as reflective surfaces which focus light in one single point was already presented 1637 by René Descartes [25]. In 1970, Rees [91] was the first one who patented the combination of a hyperbolic mirror with a perspective camera. In the last decades, catadioptric camera systems have gained wide popularity in the robotic community ([112, 11]). These camera systems are able to establish a 360° horizontal field of view and a vertical field of view larger than 60° with a very flexible geometry as the shape of the reflecting surface is a powerful design factor. For instance, the vertical field of view and the spatial resolution depend on the type and the parameters of the reflecting surface.

Fig. 1.2a shows the large field of view by capturing a traffic scene with a catadioptric camera. In comparison, Fig. 1.2b depicts the limited field of view of an image obtained by capturing the same scene with a perspective camera. The catadioptric image can be unwarped to the most popular cylindrical panoramic view representation as shown in Fig. 1.2c. However, this unwarped panoramic view is used for user convenience as visualization only. The red boxes in the catadioptric and in the unwarped panoramic image show the boundary of the field of view of the perspective camera. Unfortunately, catadioptric images have a lower spatial resolution with the same camera due to the fact that a much bigger field of view is mapped on the camera sensor and the circular image is pictured on a rectangular imager which leads to the black margin. Besides, the images suffer from distortions and defocussing blur caused by the use of a curved reflector.

In the robotic research community, catadioptric cameras are very popular for monoscopic applications. They are often used for robotics indoor perception and navigation [123], particularly in the field of the *RoboCup* [121], as well as in outdoor localization and ego-motion estimation tasks [82, 47], and in the area of video surveillance [83, 85]. Furthermore, there are some approaches which have considered catadioptric cameras for applications in driver assistance systems or autonomous vehicles. Catadioptric cameras are used for lane tracking functions [22], monoscopic visual odometry [95], as well as capturing the complete environment of the vehicle. Gandhi and Trivedi propose to use one or multiple catadioptric cameras to analyze the surrounding outside of a vehicle [37, 38] and inside the vehicle [58, 119]. In [29], Ehlgen et al. use two catadioptric cameras as a backing-up aid system by remapping the images in a birds-eye view image to provide an intuitive overview of the scene. An extension to this system with four catadioptric cameras for eliminating blind spots for trucks is given in [28].

Omnidirectional Stereo Vision At least two camera images (binocular) are necessary to compute 3D information. Stereo vision with two perspective cameras, similar to the distance estimation of humans, is a well established research field. In case of omnidirectional images several configurations are possible to build up a stereoscopic camera system, namely

- with multiple catadioptric cameras,
- with omnidirectional images generated by mosaicing techniques, or
- with a catadioptric camera and another sensor (e.g., a perspective camera or a laser scanner).

In the approach of Zhu [130] different configurations of omnidirectional stereo setups with multiple omnidirectional cameras or images captured with a mosaicing technique are discussed with respect to the numbers and configuration of viewpoints. Concerning the stereoscopic setup with two catadioptric cameras, Gluckman et al. [48] present a compact panoramic stereo camera system with two catadioptric cameras with hyperbolic mirrors on top of each other for vertically aligned stereo vision. This configuration allows a very simple epipolar geometry. In [74] such a vertically aligned stereo system is used for object detection and for monitoring blind spots of vehicles. Many approaches construct a vertically aligned imaging system with only one single camera to reduce the calibration process. Some authors [17, 31] propose to construct a double lobed mirror for a single camera catadioptric stereo vision system. Jang et al. [60] use two hyperbolic mirrors to achieve catadioptric stereo with only one lens and in [125] a single camera is used in combination with a mirror and a concave lens to improve the accuracy of the 3D reconstruction. However, single camera omnidirectional stereo systems have only a small spatial image resolution due to the fact that two complete scene images are captured with one sensor. Moreover, vertically aligned stereo systems with double lobed or two separate mirrors have the disadvantage that the baseline is very short and accurate reconstruction is possible in a short range only.

Two horizontally aligned catadioptric cameras avoid the problem of a short baseline between the cameras, especially, when the cameras are mounted on the left and right side of a vehicle. However, the accuracy of the 3D reconstruction with two horizontally aligned cameras depends on the azimuth angle of the 3D point and therefore varies for different positions as shown in [105]. Sogo et al. [108] propose to use N-occular stereo (multiple catadioptric cameras) to compensate the observation error for a human tracking application. In [40] two horizontally aligned catadioptric cameras are mounted on the left and right of the rearview mirror for a large field of view stereo vision algorithm. However, the 3D measurements are not very accurate and the images are reconstructed to multiple virtual perspective images to apply standard stereo vision algorithms. In [37] and [29] two catadioptric sensors are also used on the left and right side of the vehicle. Yet, both approaches do not compute 3D information but rather give the driver a visualization of the entire environment.

Another approach to recover 3D information is by taking panoramic mosaics computed from a rotating camera at different locations [59, 64]. In [57] a perspective stereo head consisting of two cameras is rotated while capturing two different panoramic mosaics which results in one panoramic disparity image. Peleg et al. [87] present a panoramic stereo image with circular projection from a single off-center rotating camera. Thus, a camera moves on a circular path and has multiple but fixed viewpoints. A dynamic omnistereo approach with variable viewpoint and baseline relation to find the optimal stereo configuration is presented in [131]. However, as already mentioned, panoramic images obtained by mosaicing techniques suffer from the complexity to stitch the images together. Thus, they are not suitable for dynamic environments.

The third possibility to achieve an omnidirectional stereo setup is the combination of different sensor types. Some authors [65, 97] propose to combine one omnidirectional camera with a laser scanner to obtain color values for the 3D points. However, the 3D reconstruction is limited to the field of view of the laser scanner which is in most cases a plane. Laser scanners with a larger field of view are very expensive compared to cameras. An imaging system that combines the advantages of a 360° field of view from a catadioptric camera and the high resolution from a conventional perspective camera is proposed by Lauer et al. [67] and Sturm [110]. They suggest a hybrid camera system combining a catadioptric and a perspective camera. In [1] such a hybrid omnidirectional pinhole sensor is used for stereo obstacle detection in a robotic environment. In [104] a hybrid camera system is compared with a horizontally aligned stereo system for applications in vehicles. Although such hybrid stereo systems in combinations with an active movable perspective camera are suitable to provide peripheral and foveal vision as in [61], they are not capable to provide 3D information of the complete environment.

1.2. Contribution

The contributions of this thesis are as follows:

- Stereoscopic catadioptric sensor setups are analyzed regarding their suitability for autonomous vehicles and their ability for 3D reconstruction based on a large baseline.

- A new projection model for real catadioptric cameras which are usually slightly non-central camera systems is proposed. Common projection models are either accurate but have high computational cost or efficient but not exact enough for stereo vision. We show that our proposed projection model which approximates non-central catadioptric cameras is accurate and efficient at the same time.

- A new calibration toolbox is developed which allows calibration using the proposed projection model and also handles different other projection models. Moreover, the calibration toolbox allows the extrinsic calibration of multiple cameras with respect to each other, which was still missing in existing calibration toolboxes for catadioptric cameras.

- Extensive evaluations concerning the calibration results show the advantages of the proposed projection model compared to the common central reference models. The evaluation is not only based on the possibly misleading reprojection error of the calibration targets as in existing evaluations but also reports end-to-end localization errors in a localization experiment. Moreover, the influence of deviations from the single viewpoint condition are analyzed.

- Feature matching strategies for catadioptric images, which are used as input for the ego-motion estimation, are analyzed. A still missing comparative study of different feature matching strategies on catadioptric images using high precision ground truth is presented.

- An ego-motion algorithm for autonomous vehicles with a catadioptric stereo camera system is presented. The proposed algorithm, based on two-frame motion, benefits from the new projection model. We show that omnidirectional cameras overcome major drawbacks of traditional perspective cameras for ego-motion estimation. From the estimated motion high fidelity top view maps of the driven path and the nearby surrounding are created. The maps are computed by accurately stitching remapped catadioptric top view images together.

- An approach to achieve dense panoramic 360° depth images resulting in dense 3D reconstructions from stereo catadioptric camera setups is proposed. The method refrains from constructing perspective images from the omnidirectional ones as an intermediate step. Death angles, which occur for two horizontally aligned cameras, are prevented by combining motion and spatial stereo. Planarity priors are introduced to achieve smooth 3D reconstructions.

1.3. Overview

This work describes a complete stereoscopic catadioptric camera system beginning with a new efficient and accurate projection and calibration model and ending with two applications for vehicles. Therefore, the thesis regards many different topics for omnidirectional vision and only a general overview of existing approaches using catadioptric cameras was given at the beginning. A detailed description of the state-of-the-art of the relevant topics is given at the beginning of each chapter.

The thesis is structured as follows: **Chapter 2** reviews the state-of-the-art projection models for catadioptric cameras and presents the novel efficient and at the same time exact projection model. **Chapter 3** describes the new developed calibration toolbox for multiple catadioptric cameras with the proposed projection model. Moreover, the chapter gives an extensive evaluation concerning the calibration results with different projection models. **Chapter 4** describes the stereoscopic catadioptric camera setup used for applications on autonomous vehicles. Furthermore, the chapter presents a new ego-motion estimation method for catadioptric stereo cameras and shows a comparison of the results against the results for ego-motion estimation with perspective cameras. Moreover, a comparative study for feature matching strategies on catadioptric images is given. **Chapter 5** describes a new approach for dense 3D reconstruction with catadioptric cameras and explains the construction of 360° panoramic disparity images. A conclusion is given in **Chapter 6**.

Chapter 2

Projection Models

Projection models describe the relationship between a 3D world point and a 2D image point. This relationship can be divided in the back projection and forward projection problem. The back projection formulation describes the relationship between a given 2D image point position and the resulting 3D ray in the world, which can be used for triangulation of a 3D point, for example. For applications based on minimization of the reprojection error in the image, the forward projection formulation is necessary. The forward projection describes the problem formulation which 2D image point position corresponds to a given 3D world point.

In this chapter, we present our novel accurate and efficient centered projection model for slightly non-central cameras. Before starting with a description of the centered projection model, we explain the single viewpoint condition and give an overview of existing projection models for central and non-central catadioptric cameras. Based on the discussion of existing projection models, we motivate the necessity of a novel projection function.

2.1. State-of-the-Art

Projection models for catadioptric cameras are more complex than for perspective cameras due to the non-linear mapping of a 3D ray via the mirror surface to a 2D image point. In the literature many different projection models developed for catadioptric cameras exist. They can be divided in central and non-central models depending on the cameras fulfill the single viewpoint condition.

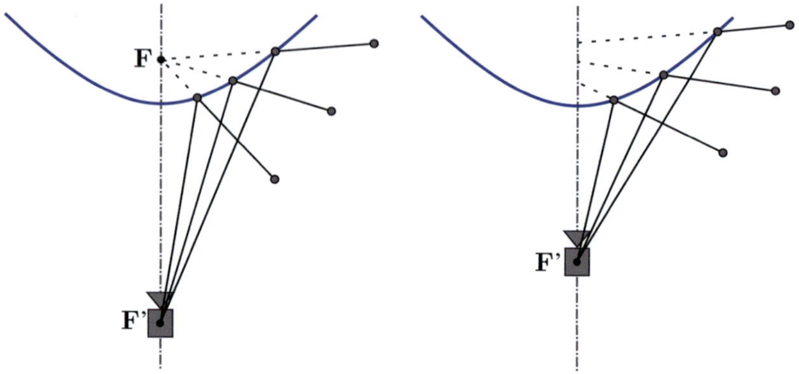

(a) Central Camera (b) Non-Central Camera

Figure 2.1.: **Central vs. Non-Central.** This figure shows the reflection rays for central cameras (a), where all reflection rays intersect in one point (**F**), and for non-central cameras (b), where the rays do not intersect in a single point.

2.1.1. Single Viewpoint Condition

Imaging systems are usually designed to closely satisfy the single viewpoint (SVP) condition. Thus, all light rays are assumed to intersect in a single effective viewpoint shown in Fig. 2.1a. Perspective cameras inherently fulfill this condition and also some catadioptric cameras are designed with the goal to fulfill this condition. Cameras that fulfill the single viewpoint condition are called central cameras. Baker and Nayar [6] describe the whole class of central catadioptric cameras which theoretically can be achieved with three combinations: a perspective lens in combination with a hyperbolic or elliptical reflector or a telecentric (orthographic) lens with a parabolic reflector. In practice, mostly hypercatadioptric systems with hyperbolic mirrors (Fig. 2.2a) and paracatadiopitric systems with parabolic mirrors (Fig. 2.2b) are used. Elliptical mirrors (Fig. 2.2c) are not suitable for panoramic vision since they capture only the upper hemisphere.

In case of a hyperbolic mirror which has two focal points, the focus of the camera is exactly placed in one focal point (**F**') while the second focal point (**F**) is the effective viewpoint where all reflection rays intersect as shown in Fig. 2.2a. For parabolic mirrors the focal point (**F**) is again the effective viewpoint. However, the camera is placed on the optical axis under the mirror on a variable distance as shown in Fig. 2.2b, since all reflected rays are parallel to the optical axis. The main advantage for cameras that have a single effective viewpoint is an easier projection function. Moreover,

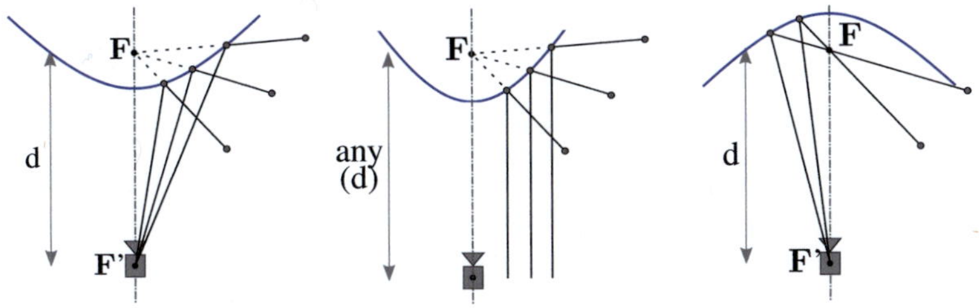

Figure 2.2.: **Single Viewpoint Cameras.** This figure shows the mirror and camera combinations which theoretically fulfill the single viewpoint condition. For the combination with a hyperbolic (a) and elliptical mirror (c) the camera is placed in the second focal point **F'** of the mirror. For a setup with a parabolic mirror (b) the camera is placed on the optical axis on a variable distance.

central cameras allow for the standard epipolar geometry and the geometrically correct remapping of a catadioptric image to other images, e.g., to perspective, panoramic, or spherical images.

Unfortunately, in practice it is nearly infeasible to fulfill the single viewpoint assumption, since perfect alignment of the camera center with the optical axis of the mirror is hardly achieved in practice. Furthermore, inaccuracies in manufacturing the mirrors as well as commonly used varifocal lenses which means the viewpoint depends non-linearly on the focus and the focal length, prevent the usage of the single viewpoint assumption. In this work we call such systems, which slightly deviate from the single viewpoint condition, quasi-central or slightly non-central cameras.

Systems where the light rays do not intersect in a single point, shown in Fig. 2.1b, are called non-central cameras. Such systems have a locus of viewpoints in three dimensions called caustic [114]. The projection models are much more complicated, mainly the forward projection, since the reflecting point on the mirror surface has an unknown position. However, non-central catadioptric cameras are more flexible in their design [115]. The position of the camera relative to the mirror is not fixed to any position and other mirror geometries, e.g., spherical or conical reflectors can be used. Moreover, special mirror designs which optimize the field of view or the image properties are possible such as specific image resolution [39], equiangular projection [20], or a distortion-free perspective projection for particular scene planes [55].

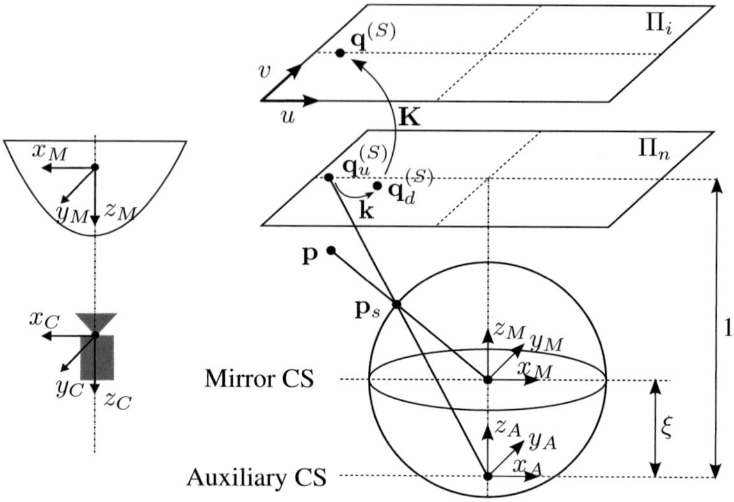

Figure 2.3.: **Sphere Camera Model.** This figure shows the axes convention on the left side and the projection model of the sphere camera model with the sphere parameters on the right side. In a first step the world point $\mathbf{p}$ is projected onto the unit sphere $\mathbf{p}_s$ and then translated and projected into a normalized plane $\mathbf{q}_u^{(S)}$. Finally, the point is mapped to an undistorted point $\mathbf{q}_d^{(S)}$ and projected to the image plane $\mathbf{q}^{(S)}$. Note that the radius of the sphere is 1, i.e., the illustration is not at scale.

2.1.2. Central Models

There exist many projection models for central catadioptric cameras. Early approaches [113, 63] focus on projection models for particular sensor types, e.g., Svoboda and Pajdla [113] propose different models for different mirror types. The most common projection model for central catadioptric systems is the sphere camera model proposed by Geyer and Daniilidis [45] and extended by Barreto and Araujo [7]. This two-step projection model unifies all central catadioptric cameras and allows for efficient forward and back projection. Ying and Hu [126] show that this model also allows the central fisheye projection. Mei and Rives [75] extend the sphere camera model with a perspective lens and add radial and tangential distortion parameters to account for misalignments between the mirror and the camera axis.

Sphere Camera Model In the following the extended sphere camera model [75] is shortly summarized. The projection is performed in five steps explained in the following and illustrated in Fig. 2.3:

1. The 3D world point $\mathbf{p}$ in the mirror coordinate system is projected onto the unit sphere with

$$\mathbf{p}_s(x_s, y_s, z_s) = \frac{\mathbf{p}}{\|\mathbf{p}\|} \tag{2.1}$$

through intersecting the sphere with the line spanned through the sphere center and the world point.

2. The point $\mathbf{p}_s$ is transferred to a new coordinate system (Auxiliary CS)

$$\mathbf{p}_\xi = (x_s, y_s, z_s + \xi)^\mathsf{T} \tag{2.2}$$

located in the focal point of the mirror.

3. The point $\mathbf{p}_\xi$ is projected into a normalized plane Π_n

$$\mathbf{q}_u^{(S)} = \begin{bmatrix} u^{(S)} \\ v^{(S)} \end{bmatrix} = \left(\frac{x_s}{z_s + \xi}, \frac{y_s}{z_s + \xi}\right)^\mathsf{T}. \tag{2.3}$$

4. The undistorted point $\mathbf{q}_u^{(S)}$ is mapped to the distorted point $\mathbf{q}_d^{(S)}$ by applying radial and tangential distortions

$$\begin{aligned} \mathbf{q}_d^{(S)} = \begin{bmatrix} u_d^{(S)} \\ v_d^{(S)} \end{bmatrix} &= (1 + k_1\rho^2 + k_2\rho^4 + k_5\rho^6)\mathbf{q}_u^{(S)} \\ &+ \begin{bmatrix} 2k_3 u^{(S)} v^{(S)} + k_4(\rho^2 + 2u^{(S)2}) \\ k_3(\rho^2 + 2v^{(S)2}) + 2k_4 u^{(S)} v^{(S)} \end{bmatrix} \end{aligned} \tag{2.4}$$

where $\rho = \sqrt{u^{(S)2} + v^{(S)2}}$ and $\mathbf{k} = [k_1, \ldots, k_5]^\mathsf{T}$ are the distortion parameters.

5. Finally, the point $\mathbf{q}_d^{(S)}$ is projected to the image point

$$\begin{bmatrix} \mathbf{q}^{(S)} \\ 1 \end{bmatrix} = \underbrace{\begin{bmatrix} f_u\eta & f_u\eta\alpha & c_u \\ 0 & f_v\eta & c_v \\ 0 & 0 & 1 \end{bmatrix}}_{\mathbf{K}} \begin{bmatrix} \mathbf{q}_d^{(S)} \\ 1 \end{bmatrix} \tag{2.5}$$

in the image plane Π_i with the intrinsic projection matrix $\mathbf{K}$ including the focal lengths $[f_u, f_v]$, the principal point $[c_u, c_v]$, and the skew factor α. The focal lengths f_i with $i \in \{u, v\}$ and the mirror parameter η cannot be determined independently and will be denoted as $\zeta_i = f_i\eta$.

The parameters ξ and η depend on the reflecting surface. A table with the parameters for different mirror types is given in [75]. Note, in difference to other projection models the *Mei* implementation does not consider the image flip and the z-axis of the camera coordinate system indicates the opposite direction (Fig. 2.3).

Polynomial Distortion Model Another common projection model is the polynomial based representation from Scaramuzza et al. [96] based on the formulation from Mičušík and Pajdla [77]. This model handles the system as a unique compact system and assumes that the catadioptric image is a highly distorted image. In the following a short summary of the polynomial model is given and Fig. 2.4 illustrates the relationship.

1. The model assumes that the projection of a 3D world point $\mathbf{p}$ onto the sensor plane $\mathbf{q}_s^{(D)}$ in metric coordinates and its image on the camera plane $\mathbf{q}^{(D)}$ in pixel coordinates are related by an affine transformation

$$\mathbf{q}_s^{(D)} = \mathbf{A}_S\mathbf{q}^{(D)} + \mathbf{t}_S \tag{2.6}$$

with

$$\mathbf{A}_S = \begin{bmatrix} c & d \\ e & 1 \end{bmatrix} \quad \text{and} \quad \mathbf{t}_S = \begin{bmatrix} c_u \\ c_v \end{bmatrix} \tag{2.7}$$

to consider small misalignments and the digitizing process.

2. The relationship between a point on the sensor plane $\mathbf{q}_s^{(D)}$ and the vector $\mathbf{p}_p$ from the viewpoint $\mathbf{F}$ through the 3D point $\mathbf{p}$ depends on the non-linear function $f^{(D)}$

$$\mathbf{p}_p = \begin{bmatrix} \mathbf{q}_s^{(D)} \\ f^{(D)}(\|\mathbf{q}_s^{(D)}\|) \end{bmatrix} \tag{2.8}$$

where

$$f^{(D)}(\nu) = d_0 + d_1\nu^1 + \ldots + d_N\nu^N \tag{2.9}$$

with $\nu = \|\mathbf{q}_s^{(D)}\|$ and $d_0, \ldots, d_N$ are the polynomial parameters.

This formulation allows particularly an efficient back projection to solve the problem which 3D ray corresponds to a given 2D image point. For the forward projection the polynomial equation needs to be addressed by finding the roots of the polynomial equation which is more time-consuming.

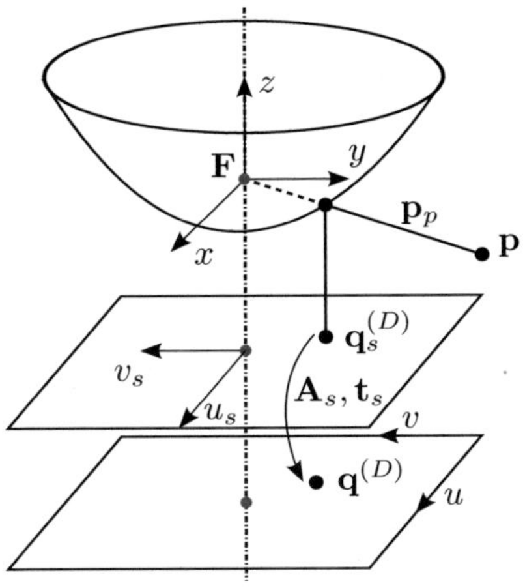

Figure 2.4.: **Polynomial Model.** This figure shows the projection model for the polynomial distortion model, where the vector $\mathbf{p_p}$ through the focal point $\mathbf{F}$ and the world point $\mathbf{p}$ depend on the non-linear function $f^{(D)}$ of the point $\mathbf{q}_s^{(D)}$. The point on the sensor plane $\mathbf{q}_s^{(D)}$ and the point on the camera plane $\mathbf{q}^{(D)}$ are related by an affine transformation.

2.1.3. Non-Central Models

There are also several works for non-central catadioptric projection models. Non-central catadioptric projection models are very accurate, but they suffer from a high computational time. The forward projection for non-central cameras is very difficult since the reflecting point on the mirror surface has an unknown position, while the back projection is much simpler [112]. Therefore, some works describe only the back projection with a generic camera model [51, 111, 90].

For the forward projection, which is necessary for applications based on minimizing the reprojection error in the image, there is no closed-form solution. Thus, some researchers [77, 70, 109, 18] propose to solve the problem to find the reflection point on the mirror surface by a computationally expensive non-linear optimization which requires an initial estimate of the pixel coordinates. Gonçalves and Nogueira [49] increase efficiency by reducing the complexity to compute the reflection point to a 1D search problem. However, they mention that they still require around 200 seconds for projecting 10 000 3D points to the image plane via a hyperbolic mirror.

Recently, Agrawal et al. [3] presented an analytical forward projection for axial non-central cameras with quadratic-shaped mirrors. In [2], they improve their analytical solution for non-axial configuration. They show that the reflecting point on the mirror surface can be obtained by solving an 8th degree polynomial equation. Compared to projection models based on optimization, they achieve a 40 times speed up. However, due to the complex root finding problem of the 8th degree polynomial, this projection model is still not real-time compatible. An overview of the analytical forward projection is given in the next section, since we use the same model as part of our base model.

In summary it can be stated that central models in general are very efficient but lack in accuracy when the camera does not exactly fulfill the single viewpoint condition. Thus, using a central projection model for a slightly non-central system leads to inaccuracies in the determination of the reflecting ray and impact the performance for accuracy sensitive tasks such as 3D reconstruction, ego-motion estimation, or localization. In contrast, non-central models are very time-consuming, on the other hand they are very precise for all types of catadioptric cameras independent of the mirror to camera placement.

2.2. Centered Projection Model

Common projection models are either efficient but rely on central models and do not consider misalignments separately or they are accurate complex non-central models which are not very efficient. In this work, we propose a novel centered projection model for slightly non-central catadioptric cameras which is accurate and at the same time efficient.

Therefore, we use the fact that the distance to world points in 3D in which we are interested is often large (> 1 m) compared to the deviations from the single viewpoint (< 10 cm) as illustrated in Fig. 2.5. This leads us to the conclusion that getting the orientation of the viewing rays correctly is more important than considering the translational deviation from the single viewpoint exactly. Unfortunately, calibrating a slightly non-central catadioptric camera system using efficient central projection models introduces a bias in the orientation of the viewing rays. This is based on the fact that for practical reasons the calibration patterns are presented in the vicinity of the camera (< 1 m). This wrong relationship between the viewing ray observation and the position on the image plane is responsible for the fact that central models perform worse for real catadioptric cameras which are not perfectly aligned. The true viewing ray (red) and a central ray after calibration (blue) are illustrated in Fig. 2.5.

Hence, we propose the centered projection method which is divided in three steps:

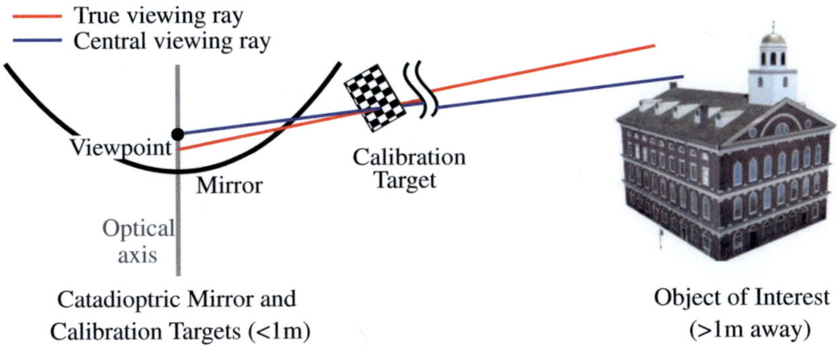

Figure 2.5.: **Viewing Ray Relationship.** This figure shows the exact viewing ray (red) and the calibrated viewing ray using a central camera projection model (blue) for a slightly non-central system with calibration targets in the vicinity of the camera. The distances to the objects of interest are usually very large compared to the deviations of the viewing rays from the single viewpoint as well as to the distances to the calibration targets.

1) Obtain the exact viewing ray orientations by using an exact non-central base model.

2) Compute an optimal single viewpoint and center the viewing rays by shifting the viewing rays to intersect the viewpoint but keep their orientation.

3) Remap the observations by projecting the centered viewing rays with a simple central projection model.

The relationship between the exact and remapped viewing rays is illustrated in Fig. 2.6. The exact viewing rays computed with the non-central base model are depicted in solid lines and the centered viewing rays are shown in dashed lines going through the optimal viewpoint (black dot). This leads to a mapping where points at infinity are projected to the same pixels as in the non-central base model and approximation accuracy gracefully degrades in the immediate vicinity of the camera center. After remapping the observations using a simple central projection model, we only use the simple and efficient central projection model whenever a projection function is required. The remapping can be pre-computed and efficiently applied to the whole image or to individual feature points similar to undistortion or rectification maps for perspective cameras.

The centered approximation is general and applicable to all slightly non-central catadioptric systems. In this work, we use a geometric model as

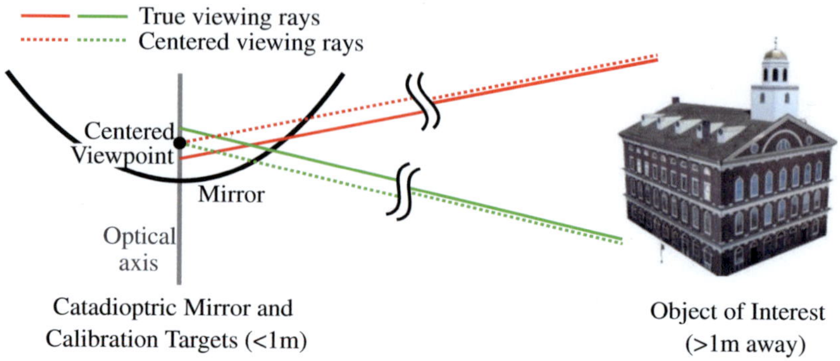

Figure 2.6.: **Centered Projection Model.** This figure shows the true viewing rays emanating from the optical system computed with a non-central base model (bold). The black dot shows the optimal centered viewpoint and the dashed lines the viewing rays of the centered approximation. Since the distance to the objects is very large the optimal viewing rays are similar to the true ones.

accurate non-central base model which needs a quadratic description of the mirror surface. For the simple central model after the centering process, which we call central-centered model, we use a more general central model based on the angle representation which needs only the viewing rays. However, the proposed idea to center slightly non-central cameras can also perform with other non-central base models as well as other central-centered models. In this chapter, we explain the non-central base projection model and the central-centered projection model which we use in this work as well as the computation of the optimal viewpoint. The calibration process and the benefits of calibrating a real catadioptric camera system with the proposed projection model in contrast to existing central models are presented in Chapter 3.2.

2.2.1. Non-Central Base Model

The non-central base model, to obtain the accurate viewing ray direction, is based on the geometric analytical forward projection model which was originally presented by Agrawal et al. [2, 3] and extended by us [106] to a complete projection model with a perspective camera including lens distortions. For the geometric model we assume a quadric mirror surface with the parameters A, B and C that can be described as

$$x_m^2 + y_m^2 + Az_m^2 + Bz_m - C = 0. \tag{2.10}$$

Mirror Type	Spherical Parameter	Parameter Conversion	Mirror Equation
Hyperboloid	$A < 0, C < 0$ $B = 0$	$A = -\frac{b^2}{a^2}$ $C = -b^2$	$\frac{z^2}{a^2} - \frac{x^2+y^2}{b^2} = 1$
Ellipsoid	$A > 0, C > 0$ $B = 0$	$A = \frac{b^2}{a^2}$ $C = b^2$	$\frac{z^2}{a^2} + \frac{x^2+y^2}{b^2} = 1$
Paraboloid	$A, C = 0$	$B = -2a$	$z = \frac{x^2+y^2}{2a}$

Table 2.1.: **Central Mirror Types.** This table provides the parametrization for all relevant mirror types that fulfill the SVP condition and the relationship between sphere parameters A, B, C (see Eq. 2.10) and manufacturing mirror parameters a, b.

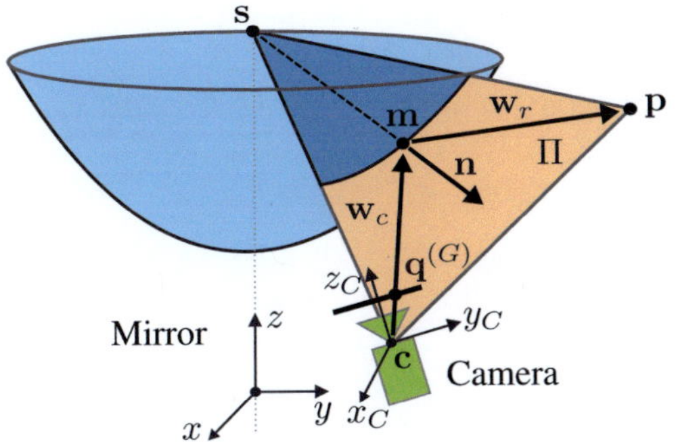

Figure 2.7.: **Geometric Projection Model.** This figure shows the non-central geometric base projection model with the parameters which we use to obtain the exact viewing rays.

This representation includes the standard parametrization for all relevant mirror types that fulfill the single viewpoint condition as shown in Table 2.1.

The geometric projection model maps a 3D world point $\mathbf{p}$ via the point of reflection $\mathbf{m} = [x_m, y_m, z_m]^T$ on the mirror surface to a pixel $\mathbf{q}^{(G)}$ on the image plane depending on the camera pose $\mathbf{c}$. This relation is depicted in Fig. 2.7. The points are represented in the mirror coordinate system if they have no index. The index C denotes the camera coordinate system. The point of reflection on the mirror surface $\mathbf{m}$ can be obtained analytically from the law of reflection and the pre-condition that the point of reflection is located on the mirror surface. This induces two constraints. The first constraint can be derived from the law of reflection

$$\mathbf{w}_r = \mathbf{w}_c - \frac{2\mathbf{n}(\mathbf{w}_c^\mathsf{T}\mathbf{n})}{\mathbf{n}^\mathsf{T}\mathbf{n}} \tag{2.11}$$

with the normal vector $\mathbf{n} = [x_m, y_m, Az_m + B/2]^\mathsf{T}$ of the mirror surface at point $\mathbf{m}$, the incoming ray $\mathbf{w}_c = (\mathbf{m} - \mathbf{c})$ and the reflecting ray $\mathbf{w}_r$ with $\mathbf{w}_r \times (\mathbf{p} - \mathbf{m}) = 0$. The second constraint can be derived from the reflection plane Π on which the world point $\mathbf{p}$, the camera $\mathbf{c}$ and reflection point $\mathbf{m}$ are located. The plane Π can be represented by $\mathbf{p}$, $\mathbf{c}$ and the intersection point between the normal vector $\mathbf{n}$ and the z-axis, which is given as $\mathbf{s} = (0, 0, z_m - Az_m - B/2)^\mathsf{T}$. The normal vector of the plane $\mathbf{n}_\Pi$ is given as

$$\mathbf{n}_\Pi = (\mathbf{p} - \mathbf{c}) \times (\mathbf{s} - \mathbf{c}). \tag{2.12}$$

Since $(\mathbf{m} - \mathbf{s})$ is orthogonal to the normal of Π, the plane can be described by

$$(\mathbf{m} - \mathbf{s})^\mathsf{T} \cdot \mathbf{n}_\Pi = 0. \tag{2.13}$$

By substituting the mirror equation (Eq. 2.10) to both constraints from Eq. 2.11 and Eq. 2.13 and combining them, we achieve a polynomial equation

$$f(z_m, A, B, C, \mathbf{c}, \mathbf{p}) = 0 \tag{2.14}$$

with the only unknown parameter z_m from the reflecting mirror point $\mathbf{m}$.

In [2], a pre-rotation is proposed to reduce the order of the resulting polynomial projection function f. It is suggested to rotate the camera location $\mathbf{c} = [x_c, y_c, z_c]^\mathsf{T}$ around the z-axis in the way that the rotated camera $\mathbf{c}_R = (0, y_{c,R}, z_{c,R})^\mathsf{T}$ aligns with the y-z-plane. Here, the index R denotes the rotated mirror coordinate system. This could be achieved with the pre-rotation matrix $\mathbf{R}_R(\mathbf{c})$

$$\mathbf{R}_R(\mathbf{c}) = \begin{bmatrix} \kappa\,(y_c + \epsilon) & -\kappa\, x_c & 0 \\ \kappa\, x_c & \kappa\,(y_c + \epsilon) & 0 \\ 0 & 0 & 1 \end{bmatrix} \tag{2.15}$$

with $\kappa = \sqrt{x_c^2 + (y_c + \epsilon)^2}$. Thus, the rotated world point $\mathbf{p}_R = \mathbf{R}_R\,\mathbf{p}$ and the rotated camera position $\mathbf{c}_R = \mathbf{R}_R\,\mathbf{c}$ are obtained. In contrast to [2], we introduce a small positive scalar ϵ which regularizes $\mathbf{R}_R$ against the identity matrix and prevents singularities in case x_c and y_c are both small.

Using the rotated points $\mathbf{p}_R$, $\mathbf{c}_R$ and $\mathbf{m}_R$ in the three equations Eq. 2.10, Eq. 2.11 and Eq. 2.13 instead of the points $\mathbf{p}$, $\mathbf{c}$ and $\mathbf{m}$ to obtain the polynomial equation, Eq. 2.14 simplifies to an 8th degree polynomial

$$\begin{aligned} f_R(z_{m,R}, A, B, C, \mathbf{c}_R, \mathbf{p}_R) = a_0(A, B, C, \mathbf{c}_R, \mathbf{p}_R) + \\ a_1(A, B, C, \mathbf{c}_R, \mathbf{p}_R) z_{m,R} + \ldots + \\ a_8(A, B, C, \mathbf{c}_R, \mathbf{p}_R) z_{m,R}^8 = 0 \end{aligned} \quad (2.16)$$

with the unknown parameter $z_{m,R}$. The coefficients of the polynomial equation $a_0, \ldots, a_8$ only depend on known parameters, namely the sphere parameters, the rotated camera position, and the rotated world point position. A detailed description how to obtain the polynomial equation and parameters is given in Appendix A.1. The roots of f_R can be computed numerically via an eigenvalue decomposition of the companion matrix resulting in $\mathbf{m}_R = (x_{m,R}, y_{m,R}, z_{m,R})^\mathsf{T}$.

After computing the rotated reflection point on the mirror surface $\mathbf{m}_R$, we project the point to the image plane. Therefore, the reflection point $\mathbf{m}_R$ is back rotated in the original mirror coordinate system and transformed into the camera coordinate system with

$$\mathbf{m}_C = (x_{m,C}, y_{m,C}, z_{m,C})^\mathsf{T} = \mathbf{R}_C \mathbf{R}_R^{-1} \mathbf{m}_R + \mathbf{t}_C \quad (2.17)$$

where $\mathbf{R}_C$ and $\mathbf{t}_C = \mathbf{c}$ denote the transformation between mirror and camera coordinates.

The normalized projection with the geometric base model $\mathbf{q}_n^{(G)}$ of the point $\mathbf{m}_C$ on the mirror surface in the camera coordinate system is given by

$$\mathbf{q}_n^{(G)} = \begin{bmatrix} x_n \\ y_n \end{bmatrix} = \begin{bmatrix} x_{m,C}/z_{m,C} \\ y_{m,C}/z_{m,C} \end{bmatrix}. \quad (2.18)$$

The distorted point $\mathbf{q}_d^{(G)}$ is computed from

$$\begin{aligned} \mathbf{q}_d^{(G)} &= (1 + k_1 r_n^2 + k_2 r_n^4 + k_5 r_n^6)\, \mathbf{q}_n^{(G)} \\ &+ \begin{bmatrix} 2k_3 x_n y_n + k_4(r_n^2 + 2x_n^2) \\ k_3(r_n^2 + 2y_n^2) + 2k_4 x_n y_n \end{bmatrix} \end{aligned} \quad (2.19)$$

with $r_n = \sqrt{x_n^2 + y_n^2}$ and $\mathbf{k} = [k_1, \ldots, k_5]^\mathsf{T}$ denote the distortion parameters.

Finally, the point $\mathbf{q}_d^{(G)}$ is projected to the image plane via

$$\begin{bmatrix} \mathbf{q}^{(G)} \\ 1 \end{bmatrix} = \underbrace{\begin{bmatrix} f_u & \alpha f_u & c_u \\ 0 & f_v & c_v \\ 0 & 0 & 1 \end{bmatrix}}_{\mathbf{K}} \cdot \begin{bmatrix} \mathbf{q}_d^{(G)} \\ 1 \end{bmatrix}. \tag{2.20}$$

Here, $\mathbf{q}^{(G)} = (u, v)^\mathsf{T}$ denotes a pixel in the image plane and $\mathbf{K}$ is the intrinsic projection matrix depending on the intrinsic parameters of the camera, the focal lengths f_u, f_v, the principal point c_u, c_v, and the skew parameter α. These intrinsic projection parameters and the distortion parameters $\mathbf{k}$ combined with the mirror parameters A, B, C and the camera location $\mathbf{c}$ and rotation $\mathbf{R}_C$ define the set of all intrinsic parameters of the non-central geometric base projection model.

2.2.2. Optimal Viewpoint

After obtaining all exact viewing rays, we compute the optimal single viewpoint $\mathbf{v}$, which is the point that is closest to all viewing rays as shown in Fig. 2.8. The set of all reflected rays can be described by

$$\{\mathbf{m} + \lambda \mathbf{w}_r(\mathbf{m}) \mid \mathbf{m} \in \mathcal{M}\} \tag{2.21}$$

where $\mathcal{M}$ is the set of all points on the mirror surface. To find the optimal viewpoint $\mathbf{v}$, we minimize the squared distance between the single viewpoint and the set of all reflected rays. The distance d is given as

$$d = \frac{\|(\mathbf{v} - \mathbf{m}) \times \mathbf{w}_r\|}{\|\mathbf{w}_r\|}. \tag{2.22}$$

With

$$(\mathbf{a} \times \mathbf{b}) \cdot (\mathbf{a} \times \mathbf{b}) = \|\mathbf{a}\|^2 \|\mathbf{b}\|^2 - (\mathbf{a} \cdot \mathbf{b})^2 \tag{2.23}$$

we compute the squared distance as

$$d^2 = \frac{\|(\mathbf{v} - \mathbf{m})\|^2 \|\mathbf{w}_r\|^2 - \left\|(\mathbf{v} - \mathbf{m})^T \mathbf{w}_r\right\|^2}{\|\mathbf{w}_r\|^2}. \tag{2.24}$$

Thus, our minimization criteria can be formalized for all rays as

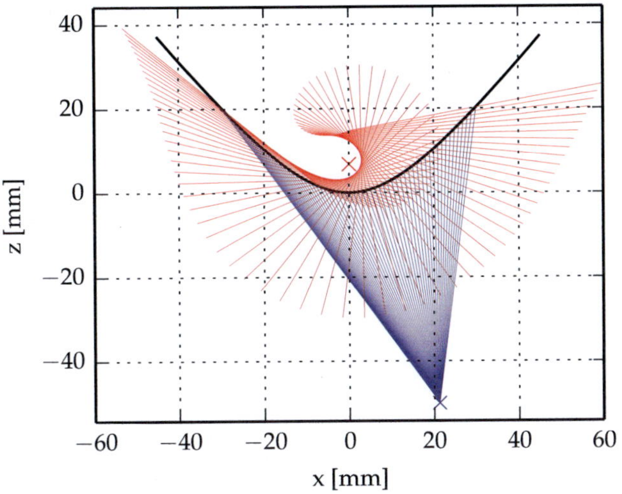

Figure 2.8.: **Optimal Viewpoint.** This figure shows the optimal viewpoint for a non-central projection where the camera location (blue cross) is laterally and axially displaced. The optimal viewpoint (red cross) is obtained by minimizing the distance to all reflected rays (red).

$$\mathbf{v} = \underset{\tilde{\mathbf{v}}}{\operatorname{argmin}} \int_{\mathcal{M}} d^2 d\mathbf{m} \tag{2.25}$$

$$= \underset{\tilde{\mathbf{v}}}{\operatorname{argmin}} \int_{\mathcal{M}} \left(\|\tilde{\mathbf{v}} - \mathbf{m}\|^2 - ([\tilde{\mathbf{v}} - \mathbf{m}]^\mathsf{T} \mathbf{w}_r(\mathbf{m}))^2 \right) \, d\mathbf{m} \tag{2.26}$$

with normalized reflected ray $\|\mathbf{w}_r\|^2 = 1$. To compute the optimal $\mathbf{v}$ that minimizes the integral, its derivative with respect to $\mathbf{v}$ should be zero.

This yields the integral equation

$$\int_{\mathcal{M}} \left(\mathbf{v} - \mathbf{m} - \mathbf{w}_r(\mathbf{m}) \left(\mathbf{v} - \mathbf{m}\right)^\mathsf{T} \mathbf{w}_r(\mathbf{m})\right) \, d\mathbf{m} = \mathbf{0} \tag{2.27}$$

which is linear in $\mathbf{v}$. This integral can be approximated to arbitrary precision by a summation over a discretized set of surface points $\mathcal{M}$. Hence, the optimal viewpoint $\mathbf{v}$ can be computed with a linear least square algorithm

$$\mathbf{b}_l = \mathbf{H}_l \mathbf{v} \tag{2.28}$$

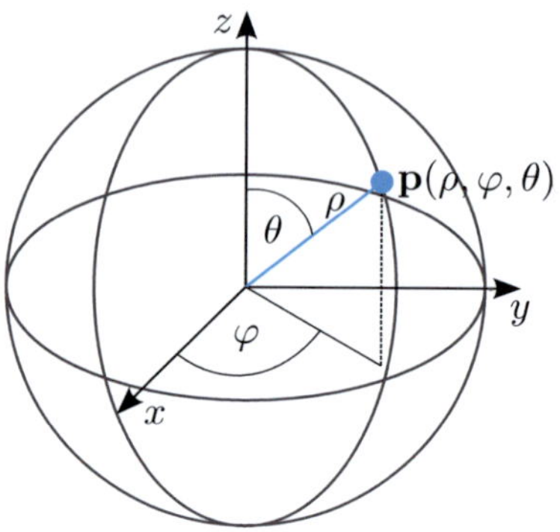

Figure 2.9.: **Spherical coordinates.** This figure shows the relationship for a 3D world point $\mathbf{p}$ between Cartesian coordinates (x, y, z) and spherical coordinates (ρ, φ, θ).

with

$$\mathbf{b}_l = -\mathbf{m} + \mathbf{w}_r \mathbf{m}^\mathsf{T} \mathbf{w}_r \tag{2.29}$$

$$\mathbf{H}_l = \mathbf{1} - \mathbf{w}_r \mathbf{w}_r^\mathsf{T}. \tag{2.30}$$

2.2.3. Central-Centered Model

After estimating the optimal viewpoint, we center the exact viewing rays and remap each pixel position with a central model. Therefore, we use a simple and efficient central model based on the angle representation which we call central-centered projection model. For this representation, the light rays through a 3D world point $\mathbf{p} = (x, y, z)^\mathsf{T}$ are only defined by the azimuth angle φ

$$\varphi(\mathbf{p}) = \arctan\left(\frac{y}{x}\right) \tag{2.31}$$

and the inclination angle θ

$$\theta(\mathbf{p}) = \arctan\left(\frac{\sqrt{x^2 + y^2}}{z}\right). \tag{2.32}$$

The relationship between spherical and Cartesian coordinates is shown in Fig. 2.9.

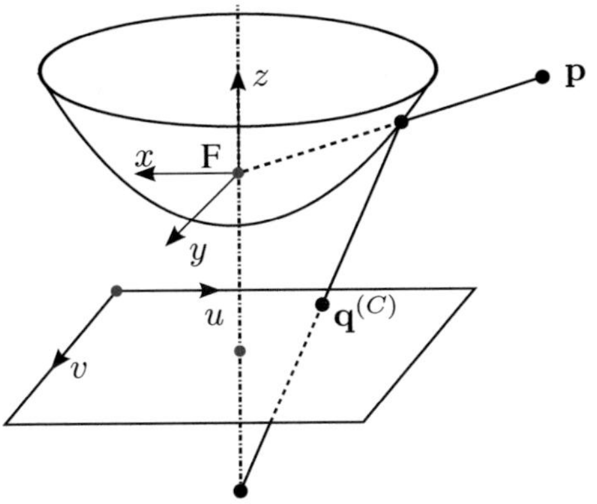

Figure 2.10.: **Central-Centered Model.** This figure shows the projection model for the central-centered model, where the world point $\mathbf{p}$ is projected to the point $\mathbf{q}^{(C)}$ on the image plane.

For the central-centered projection model (Fig. 2.10), the image point $\mathbf{q}^{(C)}$ corresponding to the world point $\mathbf{p}(\rho, \varphi, \theta)$ is defined as

$$\mathbf{q}^{(C)} = \begin{bmatrix} u \\ v \end{bmatrix} = Q(\varphi, \theta)|_{c_u, c_v, \mathbf{b}} = \begin{bmatrix} c_u \\ c_v \end{bmatrix} + \sum_{i=0}^{k} b_i \theta^i \begin{bmatrix} \cos\varphi \\ \sin\varphi \end{bmatrix} \tag{2.33}$$

where c_u, c_v are the location of the image center, $\mathbf{b} = (b_1, \ldots, b_k)^\mathsf{T}$ are the polynomial coefficients which describe the relationship between the angle θ and the distance from the image center, and $k \geq 1$ denotes the polynomial order.

After centering the exact viewing rays through the optimal viewpoint and remapping them with the central-centered projection model, we only use this central-centered projection model any time we use the proposed centered projection function. Note, the central-centered model exactly represents all other central models by remapping the image points (see Appendix A.2). Therefore, the polynomial order k is chosen to obtain a sufficient vertical field of view. The quality of the approximation is independent from order k which only impacts the smoothness of the displacement field shown in Chapter 3.2.3. Furthermore, the central-centered model maps 3D world points at infinity, estimated with any non-central base model exactly, i.e., equivalent to the projection it is derived from (see Appendix A.2).

This projection function $Q(\varphi, \theta)$ describes a very efficient forward projection from a known 3D point to an image point. For the inverse back projection $Q^{-1}(u, v)$, to compute the viewing ray to a corresponding 2D image position, we find the roots of the polynomial equation

$$\begin{bmatrix} (c_u - u)/\cos\varphi \\ (c_v - v)/\sin\varphi \end{bmatrix} + \sum_{i=0}^{k} b_i \theta^i = \mathbf{0} \tag{2.34}$$

with

$$\varphi = \arctan \frac{v}{u} \,. \tag{2.35}$$

Compared to the forward projection, the inverse projection is a rather slow, similar to the forward projection step of the *Scaramuzza* model.

Chapter 3

Calibration

A calibration process is required to estimate the intrinsic and extrinsic parameters of a projection model, before performing any application with the camera setup which needs metric scene measurements. The calibration task is very import since the accuracy of further applications depends mainly on the accuracy of the calibration result.

In this work, we develop a novel catadioptric stereo camera calibration toolbox for multiple quasi-central catadioptric cameras. The toolbox allows the calibration of camera parameters with the proposed centered projection model and can also handle other projection models. Before we explain the calibration toolbox, we review existing methods to calibrate omnidirectional cameras. In the end of this chapter, we evaluate the presented calibration process involving the proposed centered projection model in simulation and in real-world experiments in comparison to state-of-the-art catadioptric calibration models.

3.1. State-of-the-Art

Many works exist which are focusing on calibrating monoscopic catadioptric cameras with various projection models and different calibration methods. Most of them consider the catadioptric camera setup as an overall system. Recently, Puig et al. [89] have given an overview of existing catadioptric calibration methods and propose a taxonomy for omnidirectional camera calibration methods which classifies calibration methods depending on the calibration technique into five categories: Line-based calibration [44, 8, 127], 2D pattern calibration [96, 75], 3D pattern calibration [88], self-calibration [77], and polarization imaging [81]. Moreover, the authors present a comparison of four open source available toolboxes: Two based on planar patterns from Scaramuzza et al. [96] and from Mei and Rives [75], one based on a 3D pattern from Puig et al. [88], and one based on lines in the image from Barreto and Araujo [8]. While former calibration comparisons analyzed only the reprojection error in the image, this comparison evaluates the calibration methods by analyzing the reprojection error and an error of a structure from motion experiment with two images. The analyzed toolboxes use different projection functions but all are based on the assumption of a central model. The authors conclude that all methods perform similar and give an accurate reconstruction result for central cameras. However, they do not consider the fact that the central catadioptric cameras do not exactly fulfill the single viewpoint condition.

There are also some approaches for calibrating non-central catadioptric cameras based on accurate but time-consuming non-central projection models. However, to the best of our knowledge there is no public available toolbox for non-central catadioptric cameras.

Concerning the calibration method, in this work we focus on planar checkerboards as calibration targets since 2D patterns are easy to employ and constrain the problem sufficiently well. Such calibration methods need several images of planar checkerboards at different unknown orientations and positions with a sufficient distribution over the complete catadioptric image. The calibration methods presented by Mei and Rives [75] and by Scaramuzza et al. [96, 98] also use planar checkerboards. Both models can cope with all kinds of central catadioptric cameras and with fisheye cameras. Both approaches assume a central camera even if the authors mention that small deviations from the single viewpoint can be handled. However, both toolboxes are only adaptable for monoscopic camera calibration and need further extensions for multiple cameras. Anyway, both calibration toolboxes are very popular and many researchers use them for further applications. In this work, we use both calibration models as reference methods. Therefore, we summarize both calibration models in the following.

The Mei Toolbox: The *Mei toolbox* [75] is based on the extended sphere camera model (see Section 2.1.2) with distortion parameters that consider real-world errors such as misalignments between the mirror and camera axis. The projection model has 17 parameters: seven extrinisic parameters to describe the position of the 3D point relative to the camera coordinate system (rotation as quaternion $\mathbf{q}_{ex} \in \mathbb{R}^4$ and translation $\mathbf{t}_{ex} \in \mathbb{R}^3$), one mirror parameter ξ, four distortion parameters $\mathbf{k}_i$ (two tangential, two radial), and five camera parameters (two spherical focal lengths ζ_i, two components for the principal point (c_u, c_v), and one skew parameter α).

After initialization, the model parameters are estimated with a non-linear minimization of the reprojection error with the Levenberg-Marquardt algorithm. For initialization, the authors assume small deviations from the theoretical model and initialize the distortion and skew parameters to zero. To obtain initial values for the other camera parameters and the 2D calibration grid parameters, some user input is necessary. For an initial principal point, the user has to select the image center and the border of the catadioptric image, and for the initial focal lengths, at least three non-radial points must be selected. Besides, the toolbox contains a semi-automatic corner extraction which only needs manually labeled four edge corners. Moreover, this toolbox is also valid for fisheye lenses and spherical mirrors.

The Scaramuzza Toolbox: The *Scaramuzza toolbox* [96] is based on the polynomial model (see Section 2.1.2). The parameters of the model are the n polynomial coefficients $\mathbf{d}_n$ and the parameters c, d, e and c_u, c_v of the affine transformation matrices $\mathbf{A}_S$ and $\mathbf{t}_S$ which account for misalignments and the image center. Moreover, for each 3D point six extrinsic parameters are necessary, a rotation vector $\mathbf{r}_{ex} \in \mathbb{R}^3$, related to the rotation matrix by the rodrigues formula, and a translation vector $\mathbf{t}_{ex} \in \mathbb{R}^3$. The authors propose a 4th order polynomial $(n = 4)$ as sufficient.

The approach assumes the catadioptric image as a highly distorted image. The parameters are estimated with a four-step linear least square minimization problem, ignoring the affine transformations. Afterwards, they are refined by a two-step non-linear least square minimization problem. In the first non-linear minimization step, the extrinsic parameters of the checkerboards are estimated by ignoring the intrinsic camera parameters. In the second non-linear step, the intrinsic parameters are optimized with the previously computed checkerboard positions. The authors suggest this two-step minimization to speed up the convergence and conclude that the two-step minimization does not affect the final results. Moreover, they point out that the toolbox requires no prior knowledge about the mirror shape. The detection of the image center and border as well as of the checkerboards performs automatically without any user interaction. In addition, the toolbox can also be applied for fisheye lenses.

In the following we denote the projection model and calibration toolbox presented by Mei and Rives the *Mei* model and *Mei* toolbox. The projection and calibration model proposed by Scaramuzza et al. is called *Scaramuzza* model and *Scaramuzza* toolbox.

3.2. Catadioptric Stereo Calibration Toolbox

Previously existing catadioptric toolboxes as explained can only handle monoscopic cameras and are only compatible with their corresponding projection function. Since we use a stereoscopic catadioptric setup with two cameras for our applications, we developed a novel catadioptric stereo calibration toolbox for multiple slightly non-central catadioptric cameras in this work. Moreover, the developed toolbox allows the usage of the presented centered projection function which is efficient and accurate at the same time.

The proposed toolbox allows simultaneously for the intrinsic and extrinsic calibration of one or more cameras and contains the implementation of different projection methods. Note, for the proposed centered projection model we need the exact non-central base model and an efficient central-centered projection model. Thus, for calibrating the centered model we estimate the parameters of the fast central-centered model as well as the parameters for the non-central base model. Different parameter sets for the geometric model which we use as non-central base model can be calibrated. The calibration result for the geometric model can be used as a complete and exact but very slow projection model or as we suggest as base function for the centered model. Furthermore, the toolbox contains the most relevant central reference calibration models ([75, 96]) to evaluate the calibration results against them. The original reference toolboxes cannot be used to compare the results, because they are designed for monoscopic camera calibration only. Therefore, we transfer the calibration models in our toolbox and extend them for more than one camera. The user selects which projection model is used for the calibration.

In general, except for the central-centered projection model, the calibration process includes three main parts: A fully automatic corner extraction, the parameter initialization, and a non-linear optimization of the parameters by minimizing the reprojection error of checkerboard corners. Before starting the calibration process multiple images of known planar checkerboards are captured at different poses. To obtain an accurate description of the complete mirror and not only of parts of the mirror, the checkerboards should cover as many parts of the catadioptric image as possible and should be distributed uniformly over the image. Similar to the *Scaramuzza* toolbox, only little user input for the initialization is necessary, which is given in form of a configuration file in the beginning.

For the centered projection model, we first run the calibration process of the non-central base model to obtain the exact viewing rays to each pixel. Afterwards, we compute the optimal viewpoint and center the viewing rays.

From the centered viewing rays, we estimate the parameters of the central-centered projection model. Finally, the observations are remapped and the efficient central-centered projection function is used. Note that any time we use the centered model after the calibration process, we only have to remap the pixel positions and apply the central-centered function.

In the following, we explain the corner extraction, the calibration of the base model and the central-centered model as well as the changes to the central reference models which are added to the toolbox.

3.2.1. Corner Extraction

The first step of the calibration process is the extraction of the checkerboard corners in all images. We use an automatic corner extraction based on the approach presented in [43] for checkerboards in perspective images. We apply this detector on two image scales and predict corners non-linearly in the association stage for a better handling of catadioptric image distortions. For the stereo calibration, the detected checkerboard corners are automatically tracked in the images of multiple cameras by sorting the corners as shown in Fig. 3.1a. In order to do so, the edge corners are represented in the polar coordinate system (φ,r) and sorted corresponding to their azimuth angle φ and radius r. The sorting algorithm constrains the position of the checkerboard in the way that all corners are visible in the image and the rotation of the checkerboard is smaller than 90°. In our calibration images this sorting always works as long as all corners were detected, otherwise the calibration image is not used for the calibration process.

3.2.2. Non-Central Base Model

For calibrating the non-central base model, we observe the non-central geometric model (see Eq. 2.16 - Eq. 2.20)

$$\mathbf{q}^{(G)} = \mathbf{K} \cdot \mathbf{q}_d^{(G)}(A, B, C, \mathbf{k}, \mathbf{c}, \mathbf{R}_C, \mathbf{p}) \tag{3.1}$$

where the image point $\mathbf{q}^{(G)}$ depends on

- three spherical mirror parameters (A, B, C),
- five intrinsic camera parameters of the calibration matrix $\mathbf{K}$ (focal lengths f_u,f_v, principal point c_u, c_v, skew parameter α),
- four distortion parameters $\mathbf{k} = [k_1, k_2, k_3, k_4]^\mathsf{T}$,
- six parameters for the camera position $\mathbf{c}$ and rotation $\mathbf{R}_C$,

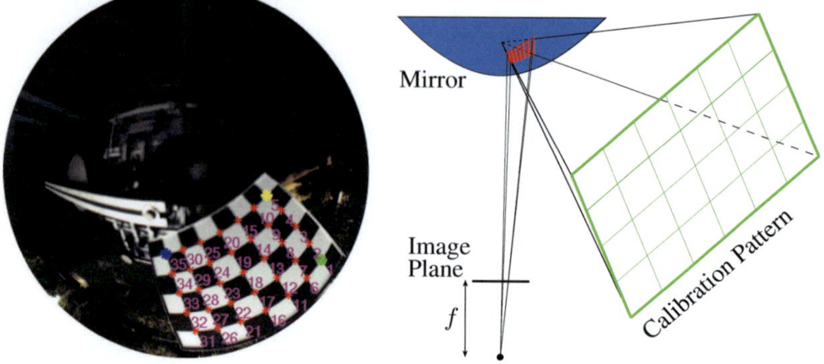

(a) Corner Extraction (b) Checkerboard Initialization

Figure 3.1.: **Calibration Initialization.** This figure (a) shows the result of the corner extraction with sorted corners, where 1 always denotes the left upper corner of the checkerboard. In (b) the relationship between the points on the mirror surface and the planar checkerboard for the checkerboard initialization is illustrated.

- and a three-dimensional world point $\mathbf{p}$.

The 3D world point $\mathbf{p}$ again depends on

- six extrinsic parameters $\mathbf{r}_{ex}$ and $\mathbf{t}_{ex}$ for each camera and
- six extrinsic parameters for each checkerboard $\mathbf{r}_{cb}$ and $\mathbf{t}_{cb}$.

For the representation of a rotation matrix $\mathbf{R}$, we use the Rodrigues formula which represents the rotation with a three-dimensional rotation vector $\mathbf{r} \in \mathbb{R}^{\mathbf{3}}$, describing the three degrees of freedom of the rotation. The camera transformations are given with respect to the mirror coordinate system which is chosen as the world coordinate system.

Initialization Before we estimate the parameters of the non-central geometric model by optimization, the model parameters and the unknown positions of the checkerboards in 3D must be initialized. For initialization, we assume the camera as a perfect central one. Hence, the camera has no distortions ($\mathbf{k} = \mathbf{0}$, $\alpha = 0$), is perfectly aligned with the mirror axis ($\mathbf{r}_c = \mathbf{0}$) and placed in the second focal point in case of a hyperbolic mirror ($\mathbf{t}_c = [0, 0, \sqrt{C/A - A}]^\mathsf{T}$). This assumption simplifies the geometric model significantly. Besides, some manual information concerning the size of the squares of the checkerboards and the initial mirror geometry parameters A, B, C as well as the radius of the mirror r_M are necessary. Moreover, the user has to select two points on an image to compute the principal point (c_u, c_v) and radius of the image r_I. The initial focal lengths

$$f_u = f_v = \frac{2\sqrt{C/A - A}}{r_M} r_I \tag{3.2}$$

can be computed with the assumption of a central model.

The initial parameters for the checkerboard positions $(\mathbf{r}_{cb}, \mathbf{t}_{cb})$ are computed with a homography [53] similar to the initialization of calibration patterns for the calibration of perspective cameras. However, the points on the image plane are not directly used to compute the homography. Instead, we use the corresponding points on the mirror surface with the approximation that the points lie on a planar plane which generates an acceptable error for initialization. Accordingly, the homography is computed between points on the mirror surface and points on the planar checkerboard as visualized in Fig. 3.1b. The points on the mirror surface are again computed from the corresponding image points with the assumption of a central geometric model. We obtain the points on the mirror surface $\mathbf{m}$ by intersecting the mirror surface with the incoming rays $\mathbf{w}_c = \mathbf{t}_c + \lambda \cdot \mathbf{q}_n^{(G)}$. This yields

$$\mathbf{m} = \mathbf{t}_c + \lambda(A, B, C, f, c_u, c_v) \cdot \mathbf{q}_n^{(G)}(f, c_u, c_v) \tag{3.3}$$

where $\mathbf{q}_n^{(G)}$ is the normalized projected image point computed with the inverse function of Eq. 2.20.

Thereby, the checkerboard positions in all cameras where the checkerboards are visible are computed. The initial camera transformation $(\mathbf{r}_{ex}, \mathbf{t}_{ex})$ between the cameras is calculated as the mean camera transformation between the checkerboards in the different cameras.

Optimization After initialization, we estimate the parameters of the complete non-central geometric model

$$\Gamma = (A, B, C, \mathbf{K}, \mathbf{k}, \mathbf{r}_c, \mathbf{t}_c, \mathbf{r}_{ex}, \mathbf{t}_{ex})^\mathsf{T} \tag{3.4}$$

with a non-linear least square algorithm minimizing the reprojection error for all n checkerboard corners and l cameras. Thus, we minimize

$$\Gamma = \underset{\tilde{\Gamma}}{\operatorname{argmin}} \sum_n \sum_l \|\mathbf{q}_{n,l}^{(E)} - \mathbf{q}_{n,l}^{(G)}(\tilde{\Gamma})\|_2^2 \tag{3.5}$$

with $\mathbf{q}^{(E)}$ the image position of the checkerboard corners estimated from the corner extraction and $\mathbf{q}^{(G)}$ the computed image position with the non-central geometric model. To run the optimization, we use the Levenberg-Marquardt algorithm [73] as implemented in MATLAB. The toolbox allows for optimization of different parameter sets of the geometric model which we evaluate in our experiments.

3.2.3. Centered Model

Depending on the calibration parameters for the non-central base model, we compute the parameters of the central-centered projection model and remap the observations to use only the efficient central-centered method in the following. Therefore, we first compute the reflecting ray to each pixel in the original catadioptric image with the non-central base model. From the reflected rays, we compute the optimal single viewpoint by solving the linear least square algorithm (see Eq. 2.28). In practice, we use equidistantly sampled rays to estimate the optimal single viewpoint. Afterwards, we center the viewing rays by shifting them to the optimal viewpoint while keeping their direction. We compute the new image position of the centered viewing rays with the central-centered model corresponding to the image points $\mathbf{q}^{(G)}$ in the original catadioptric image computed with the geometric model, which yields a residual displacement field.

We estimate the parameters of the centered model $(c_u, c_v, \mathbf{b})$ with a non-linear least square minimization. Therefore, we compute the viewing ray parameters φ and θ to each pixel in the original image from the calibrated geometric non-central camera model and minimize

$$c_u, c_v, \mathbf{b} = \underset{\tilde{c}_u, \tilde{c}_v, \tilde{\mathbf{b}}}{\operatorname{argmin}} \sum_{\varphi,\theta} \|\mathbf{q}_{\varphi,\theta}^{(G)} - \mathbf{q}^{(C)}(\varphi, \theta; \tilde{c}_u, \tilde{c}_v, \tilde{\mathbf{b}})\|_2^2, \tag{3.6}$$

the error between the original image point $\mathbf{q}^{(G)}$ and the computed image point $\mathbf{q}^{(C)}$ with the central-centered projection model (see Eq. 2.33).

The image residual $\mathbf{q}^{(G)} - \mathbf{q}^{(C)}$ after optimization defines the residual displacement field which is applied to the original image observations before using the central-centered model. The degree of the polynomial equation (Eq. 2.33) from the central-centered projection model does not impact the quality of the approximation but only affects the smoothness of the residual displacement field. In Fig. 3.2 displacement fields for polynomial equations of differing order are illustrated. In the top row, the displacement fields for remapping the observations of a central camera with the central-centered model, which were originally described with the geometric model, are shown. In the bottom row, the displacement fields for remapping a non-central camera, where the position deviates 5 mm in axial and lateral di-

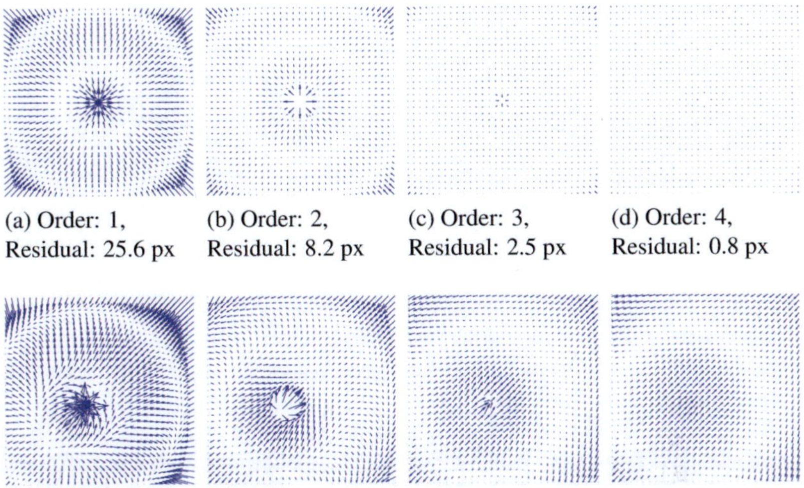

(a) Order: 1, Residual: 25.6 px
(b) Order: 2, Residual: 8.2 px
(c) Order: 3, Residual: 2.5 px
(d) Order: 4, Residual: 0.8 px

(e) Order: 1, Residual: 39.8 px
(f) Order: 2, Residual: 24.3 px
(g) Order: 3, Residual: 20.3 px
(h) Order: 4, Residual: 20.3 px

Figure 3.2.: **Displacement Fields.** This figure shows the residual displacement fields for remapping the observations with the central-centered model for polynomials of different order ($k = [1, \ldots, 4]$). The residual value below the image state the mean residual value of the image. Top row: Remapping of a central camera. Bottom: Remapping of a non-central camera with 5 mm deviation from the single viewpoint in axial and lateral direction.

rection, are depicted. Experimentally, we found a third order polynomial ($k = 3$) sufficient for providing smooth displacement fields. For successful convergence, we initialize the parameters c_u, c_v to half of the image size and the polynomial coefficients $\mathbf{b}$ to zero.

After estimating the central-centered projection parameters, the displacement field can be densely precomputed once. This remapping is similar to the undistortion task for perspective cameras. When using the centered projection model in a first step all observations are mapped by the precomputed displacement field before the central-centered projection function is applied. In practice, we stored only every fourth point in the displacement map and use cubic interpolation for remapping the observations. Note that the centered model is exact for all points at infinity and all central models (see Appendix A.2).

3.2.4. Reference Models

As already mentioned, the *Mei* and *Scaramuzza* models are added to the toolbox as common reference calibration models. Moreover, we add a third method, called *Geyer* method [45], which is the sphere camera model as used for the *Mei* toolbox without distortion parameters. Since we cannot use the original toolboxes, which are designed for monoscopic camera calibration only, some extensions are necessary. Therefore, we add the extrinsic parameters for multiple cameras to the optimization step. For simplification, we use the corner extraction from our toolbox likewise for the reference models, since the *Mei* toolbox does not provide a completely automatic corner extraction and the automatic corner extraction from *Scaramuzza* does not work reliable in our images, which was also observed in the comparison of Puig et al. [89].

Concerning the *Mei* and *Geyer* methods, only small additional changes are necessary. Instead of using the original checkerboard initialization, we use the homography based initialization for the checkerboards extrinsics and camera extrinsics which does not affect the calibration result. Hence, we use the same formulation of the rotation parameters by a 3D rotation vector as in our models instead of quaternions. For comparability, we remap the direction of the *Mei* coordinate system to the axes directions of our coordinate system. We take the focal lengths and principal point from the initialization with the central geometric model to initialize the sphere model parameter ξ and the spherical focal lengths ζ_i. For the optimization step, we only add the transformation between the cameras to the projection function.

For the *Scaramuzza* model, we propose larger changes to improve the calibration result and run a fair comparison between the different projection functions. By applying the original two-step method and the original initialization from Scaramuzza et al. to our setup, we achieve larger errors. We observe that the method is sensitive with respect to the initialization when deviating from the single viewpoint condition. We found the reason for this to be mainly numerical instabilities which can be mitigated by normalizing the polynomial coefficients appropriately. For the evaluation, we regard the original *Scaramuzza* model and an improved *Scaramuzza* model. For both models, we add the extrinsic camera parameters and change the optimization to optimize all parameters together in one single optimization run. For the original model, we use the original initialization proposed by the authors. However, for the improved model, we again use the homography based initialization of the checkerboards and the transformation between the cameras. The polynomial parameters are initialized with a linear least square algorithm.

For all three central reference models the optimization function is similar to the one from the non-central base model. We minimize the reprojection error of the checkerboard corners (see Eq. 3.5) with a parameter list Γ corresponding to the projection models.

3.3. Evaluation

In this chapter, we evaluate the calibration results for the proposed centered projection model calibrated with the presented catadioptric stereo calibration toolbox. The proposed model is evaluated against the popular central calibration models from *Scaramuzza*, *Mei* and *Geyer* which are also added to the calibration toolbox. The advantage of the centered projection function for real catadioptric cameras which are usually slightly non-central is shown in simulation and on real-world experiments. The evaluation does not only rely on the quality of the reprojection error from the checkerboards but also on a localization experiment. Moreover, we evaluate the approximation accuracy of the centered model and the runtimes of the different projection models.

3.3.1. Sensor Setup

For the evaluation of the calibration and projection models, we use a catadioptric stereo camera system consisting of two catadioptric cameras with hyperbolic mirrors (hypercatadioptric cameras) mounted on top of the experimental vehicle. The cameras are mounted similarly to the camera setup which we later use for the real-world applications. We use the mirror type VS-C450 with a varifocal lens provided by *ITRobotics*. The hypercatadioptric camera system is shown in Fig. 3.3. The mirror has a vertical field of view of $75°$ (upper side $+15°$, lower side $-60°$) and the hyperbolic parameters are $a = 20.8485$ mm and $b = 26.8578$ mm. The black needle in the middle of the camera system prevents internal reflections. We use *PointGrey Flea2* color cameras with an image resolution of 5 Megapixels.

Our catadioptric camera systems are slightly non-central cameras. Experiments show the deviation from the optimal position by approximately 20 mm in axial and 1 mm in lateral direction. In Fig. 3.3 both, the optimal viewpoint in red and the real deviated camera focal point in orange, are shown. For a catadioptric camera which perfectly fulfills the single viewpoint condition the distance between the camera and mirror or the varifocal lens should be adjusted in the way that the real and theoretical viewpoint

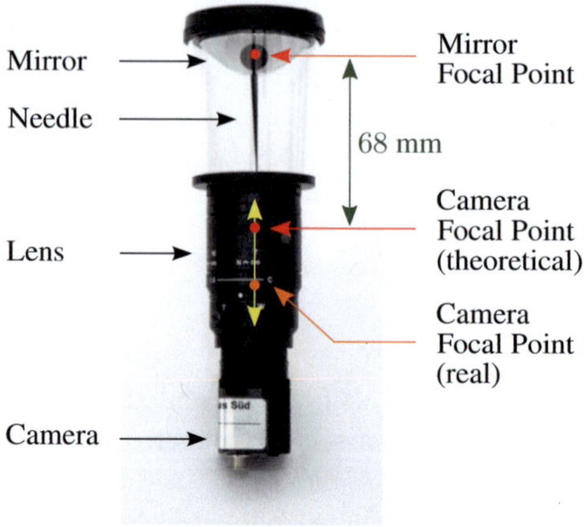

Figure 3.3.: **Slightly Non-Central Hypercatadioptric Camera.** This figure shows the hypercatadioptric camera system which we use for the calibration experiments and further real-world applications. The red dot depicts where the focal point of the camera should be to fulfill the single viewpoint condition and the orange dot shows the approximatively determined position of the focal point in our slightly non-central cameras.

coincide. The 3D world points are represented in the mirror coordinate system located between the mirror and camera with the z-axis along the positive optical axis.

3.3.2. Camera Calibration

For the experiments in simulation and with real-world data we calibrate our sensor setup with 67 calibration images each with one calibration pattern in the image. The reconstructed world positions of all calibration patterns are shown in Fig. 3.4. We use the toolbox to calibrate

- two central geometric models with different parameter sets (1-2),
- four non-central geometric models with different parameter sets (3-6),
- the proposed centered model (7), and
- four central reference models, improved *Scaramuzza* (8), original *Scaramuzza* (9), *Mei* (10), and *Geyer* (11).

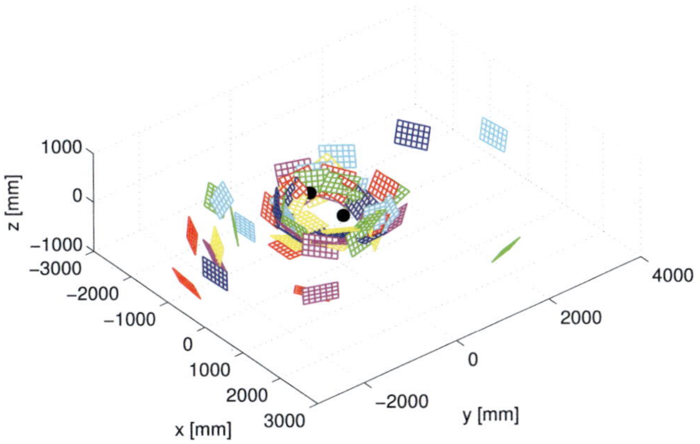

Figure 3.4.: **Calibration Pattern Positions.** This figure shows the 3D position of the 67 calibration targets in different colors around the stereoscopic camera setup, denoted as the two black dots.

The central geometric model is similar to the non-central geometric model, with the difference of a fixed camera position and fixed mirror parameters. In case of the central and non-central geometric models different parameter sets are optimized. For every geometric model the focal lengths (f_u, f_v), the principal point (c_u, c_v), and the extrinsics of the camera $(\mathbf{r}_{ex}, \mathbf{t}_{ex})$ are calibrated. The distortion parameters $(\mathbf{k})$, the camera location $(\mathbf{r}_c, \mathbf{t}_c)$, and the mirror parameters (A,B,C) are only optimized if indicated. In Table 3.1 an overview of the different projection models and the related optimized parameters are given. The numbers of the projection models given in the numeration above are the same as in the table.

As a first indicator to evaluate the quality of the projection model and the corresponding calibration result, we use the reprojection error

$$e_c = \|\mathbf{q}^{(E)} - \mathbf{q}^{(*)}\| \tag{3.7}$$

between the detected $\mathbf{q}^{(E)}$ and estimated $\mathbf{q}^{(*)}$ corners of the checkerboards. For all projection models except the centered model this is the value which is minimized during the calibration process. Note, the centered model uses a calibrated non-central base model and minimizes the residual between the original and remapped pixels. In Fig. 3.5 a boxplot shows the remaining reprojection error over all calibration images after the calibration process for the different projection models which use the minimization of the reprojection error of the corners to achieve a calibration results.

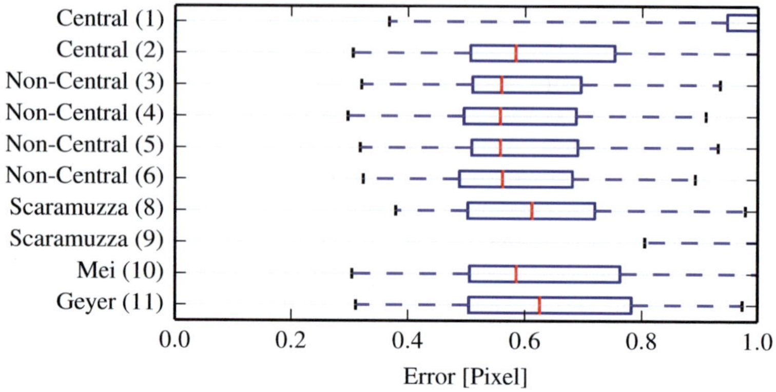

Figure 3.5.: **Reprojection Error Checkerboards.** The figure shows a boxplot representing the reprojection error e_c in pixels of the different projection models after the calibration process. The smallest reprojection error is achieved with the non-central models. The numbers of the projection models are the same as given in Table 3.1.

The smallest mean reprojection error for the corners after the calibration process, also shown in Table 3.1, is achieved with the entire non-central geometric model (6) which optimizes the mirror parameters (A,B,C), the distortions $(\mathbf{k})$, and the camera location $(\mathbf{r}_c, \mathbf{t}_c)$ except the z-component. Estimating the z-component and the mirror parameters together is not reasonable, since the mirror parameters comprise the distance between camera and mirror, which is equivalent to the z-position of the camera.

The calibration result of the centered model are not reviewed on the criteria based on the reprojection error of the checkerboard corners which is very large since the centered model is optimized for a different distance than the distance of the calibration patterns. The distance to the calibration targets with around 1 m is much smaller than the distance to the objects of interest for applications after the calibration process. Hence, the model parameters of the other projection models are optimized for a different distance than the distance of interest. To represent the distance of interest during the calibration process, the checkerboards have to be far away and therefore should be very large which makes the calibration impractical. Thus, only regarding the reprojection error of the checkerboard corners is not a sufficient criteria to evaluate the projection models and corresponding calibration results. A second reason why the reprojection error is not a sufficient value is over-

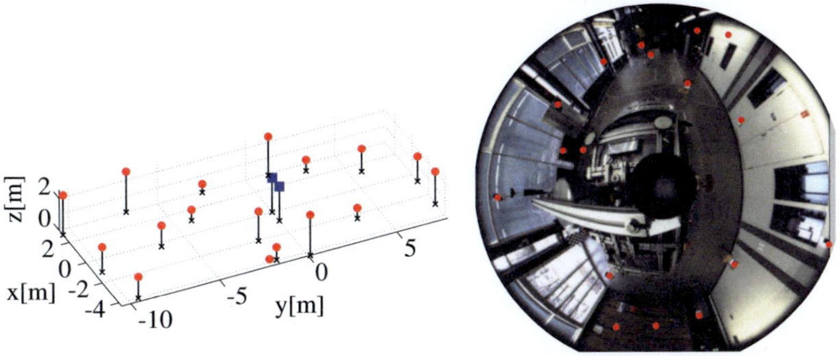

(a) Localization Setup (b) Captured Camera Image

Figure 3.6.: **Localization Experiment.** In (a) the setup for the localization experiment with the position of the landmarks (red circles), the footpoints of the landmarks and cameras (black crosses), and the camera positions (blue) are shown. In (b) the captured camera image of the scene with the detected landmarks (red) is depicted.

fitting which means the calibration reprojection error can be reduced by adding more parameters while the unobserved real-world error increases, which we show in our localization experiments.

3.3.3. Localization Experiment

We perform a localization experiment to evaluate the calibration results beyond evaluating the reprojection error. In the following, we explain the localization setup and experiment before we discuss the results in simulation and real-world experiments.

Localization Setup We perform a localization experiment where the camera positions are estimated and we accurately know the ground truth positions. We build up a localization setup consisting of 17 landmarks (red circles in Fig. 3.6a) around the stereoscopic camera setup (marked in blue). The landmarks are mounted on various heights (0 - 2.5 m) and various distances from the cameras (2.5 - 10 m). We measure the distances between all pairwise combinations of cameras and landmarks and their height over ground level (black line in Fig. 3.6a) with a high precision laser range finder. Hence, we can accurately estimate the 3D ground truth positions of all landmarks around the camera setup. We compute the 3D positions of the cameras and landmarks by minimizing all distance and height errors between the measured and computed distances with non-linear least squares, after manually initializing the positions. As landmarks we use printed 2×2 checkerboards randomly distributed in several positions in a room. The landmarks are accurately detected with the same corner detector as used for the corner extraction in the calibration toolbox. An image captured with the catadioptric camera containing the landmarks is depicted in Fig. 3.6b.

Localization Experiment Within this setup, we select 29 non-collinear landmark triplets as minimum sets for localization of the two cameras using the different projection methods. We localize our camera position by minimizing the reprojection error between the three detected landmark image positions $\mathbf{q}^{(E)}$ and the calculated points $\mathbf{q}^{(*)}$ with the corresponding model. We use a non-linear optimization similar to the optimization of the calibration parameters in Eq. 3.5 while only optimizing the extrinsic camera parameters. This yields

$$\mathbf{r}_{ex}, \mathbf{t}_{ex} = \underset{\tilde{\mathbf{r}}_{ex}, \tilde{\mathbf{t}}_{ex}}{\operatorname{argmin}} \sum_{l} \sum_{i=1}^{3} \left\| \mathbf{q}_{l,i}^{(E)} - \mathbf{q}_{l,i}^{(*)}(\tilde{\mathbf{r}}_{ex}, \tilde{\mathbf{t}}_{ex}) \right\|_2^2 \tag{3.8}$$

where the camera extrinsics $\mathbf{r}_{ex}, \mathbf{t}_{ex}$ denote the position of the cameras. We perform this localization experiment using both, a monocular (one camera, $l = 1$) and a stereoscopic (two cameras, $l = 2$) setup. For evaluating the calibration results, we consider the mean localization performance over all 29 landmark triplets for all methods.

3.3.3.1. Non Single Viewpoint Simulation

The sensitivity of projection models to deviations from the single viewpoint condition in axial and lateral direction is shown in simulation, since the pure effect caused by the deviations is hard to observe in real environment experiments. To validate the results for the single viewpoint deviation, we observe the reprojection error of the checkerboards corners after the calibration process as well as the error of the localization experiment.

We simulate a set of scenarios assuming the non-central geometric model (3) as exact ground truth model. We use the intrinsic camera parameters and extrinsic checkerboard parameters from the calibrated central geometric model (1). The mirror parameters are defined to perfect hyperbolic parameters. The only deviation from the single viewpoint assumption is the camera position which we vary in axial direction along the mirror axis

(± 20 mm in z-direction) and in lateral direction (± 10 mm in x-direction). We use an increment of 1 mm to achieve sets only distinguished by the camera position.

For every setting, we project the 3D points from the calibration patterns as well as from the localization landmarks to the image plane with the non-central geometric model selected as ground truth model. Thus, we achieve new simulated checkerboard corners and localization landmarks. Afterwards, we perform the calibration and the monoscopic localization experiment for the different projection models with the simulated calibration corners and landmark positions in the images.

Hence, for every set with a different camera position we obtain a calibration and localization result. In Fig. 3.7 the reprojection error of the checkerboards after calibration in pixels (top) and the mean monoscopic localization error in millimeters (bottom) are depicted. The errors are shown for different deviations from the single viewpoint condition in axial direction (left) and lateral direction (right). As expected, the central models without any distortion term (Central (1) and *Geyer* (11)) perform poorly for any deviation from the single viewpoint in axial or lateral direction. The simulation results show that mainly axial deviation can by adjusted with distortion parameters (Central (2) and *Mei* (10)). However, lateral displacements impact the performance much stronger. The affine transformation from the improved *Scaramuzza* model (8) also adjusts axial deviation well but cannot handle any lateral deviation.

Obviously, the localization error shows that the proposed centered model (7) handles deviations in axial and lateral direction much better than the central reference models. The fact that the reprojection errors of the checkerboards of the centered model are larger than those of the other models can be attributed to the proximity of the calibration patterns to the camera (< 1 m) compared to the localization landmarks (> 2.5 m). However, this does not effect the performance of the proposed centered model where the points of interest are more than 2.5 meters away. Here, it is clearly visible that small reprojection errors are not a sufficient indicator for a well calibrated camera with respect to some target application, e.g., localization. This confirms the assumption for the centered projection model of getting the right viewing ray direction is much more important since small orientation errors propagate to large translation errors at distance.

3.3.3.2. Real-World Experiments

We perform the same localization experiment with real-world data as described in Section 3.3.3 with both, a monoscopic and stereoscopic camera

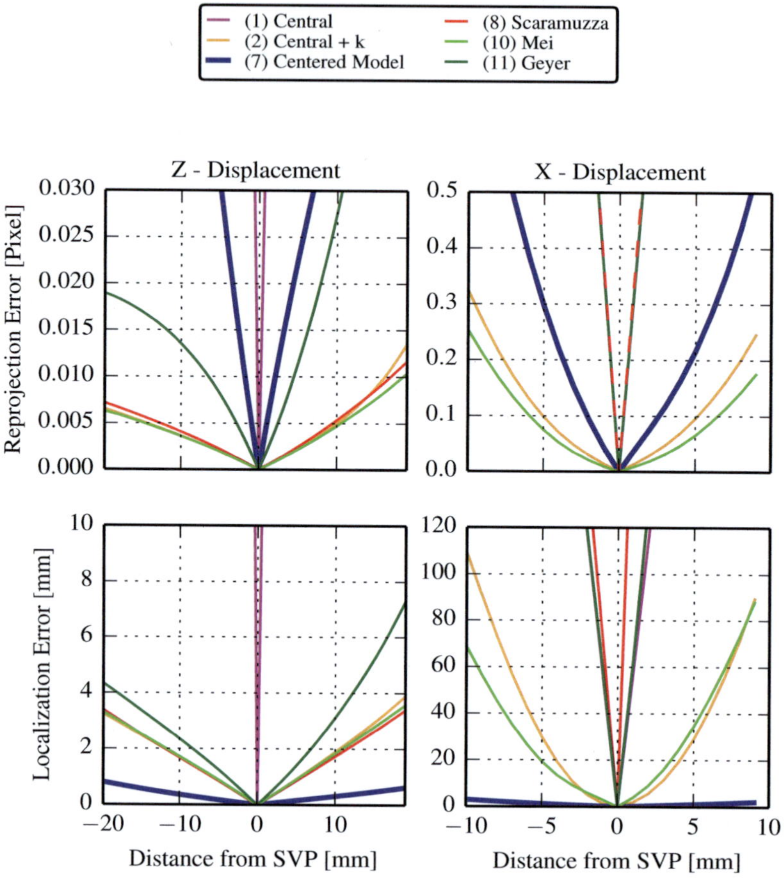

Figure 3.7.: **Simulated Displacements from the Single Viewpoint Condition.** This figure shows the reprojection errors of the checkerboard corners after calibration (top row) and the mean localization errors over all landmark triplets from the localization experiment (bottom row) when displacing the camera center axially (left) and laterally (right).

No	Method	Parameter	Repro. Error [Pixels]	Localization Error Mono [mm]	Localization Error Stereo [mm]
(1)	Geometric		1.5944	207.23	166.83
(2)	Central Model	$\mathbf{k}$	0.6241	50.89	45.56
(3)		$\mathbf{r}_c$,$\mathbf{t}_c$	0.5989	42.11	36.21
(4)	Geometric Non-	$\mathbf{r}_c$,$\mathbf{t}_c$,$\mathbf{k}$	0.5864	**40.04**	**34.64**
(5)	Central Model	$\mathbf{r}_c$,$\mathbf{t}_c$,A,B,C	0.5977	43.39	38.15
(6)		$\mathbf{r}_c$,$\mathbf{t}_c$,A,B,C,$\mathbf{k}$	**0.5850**	89.31	86.43
(7)	Centered Model		-	42.14	36.26
(8)	Scaramuzza	Improved	0.6241	49.51	48.08
(9)		Original	3.4143	771.93	687.86
(10)	Mei		0.6229	50.48	44.78
(11)	Geyer		0.6421	127.45	122.17

Table 3.1.: **Calibration and Localization Experiments.** This table shows our experiments on real data in terms of the reprojection errors of the checkerboard corners after calibration and the monoscopic as well as stereoscopic localization errors, averaged over all triplets. Moreover, the table shows the parameters which are optimized for the different geometric central and non-central models mentioned in Chapter 3.3.2.

setup. In the real-world experiments we can also evaluate the non-central geometric model which was chosen as ground truth model for the simulation. Table 3.1 shows the mean localization error of the triplets for the localization experiment with the monocular and stereoscopic setup for the different projection models and various parameter sets. Moreover, the table shows the remaining reprojection error for the checkerboards after the calibration process. An overview of the localization error for all landmark triplets and projection models for the monoscopic and stereoscopic localization is given in Fig. 3.10 and Fig. 3.11. In both figures each diagram shows the error (y-axis) for each of the 29 landmark triplets (x-axis). The rows show the different projection models while the columns depict the error in x-, y-, z-direction and the absolute error in meters. The numbers of the methods are the same as given in Table 3.1 and mentioned in Chapter 3.3.2.

The best localization performance is achieved using the non-central geometric model. In particular, model (4) with optimized camera location

$(\mathbf{r}_c, \mathbf{t}_c)$ and distortion parameters $(\mathbf{k})$ yields the smallest localization error. As expected, the non-central geometric model (6) which achieved the smallest calibration reprojection error performs worse for the localization than the other non-central models. Experiments show that minimizing the mirror parameters (A, B, C) or the distortion parameters $(\mathbf{k})$ leads to similar results but optimizing both together induces overfitting.

Due to the deviation from the single viewpoint in axial direction, the central models without any distortion model, namely the geometric central model (1) and the *Geyer* model (11), completely fail in computing the camera positions. Introducing distortion parameters to the models, as for the central geometric model (2) or the *Mei* model (10), adjusts the deviations and improves the result for the camera localization and therefore provides a better calibration result. This confirms the results obtained from simulation. As already mentioned, the original *Scaramuzza* model (9) fails completely due to numerical instabilities for non-negligible deviations from the single viewpoint condition, while the improved *Scaramuzza* model (8) performs similar to the *Mei* model (10) or the geometric model with distortions (2).

In fact, the models adjusting the deviations by some parameters, such as the central model with distortions (2) or the central reference models from *Scaramuzza* (8) and *Mei* (10) perform better than central models without any deviation parameter but worse than the non-central geometric models. Accordingly, the localization result is improved around 10 mm with the non-central models. More precisely, the non-central model reduces the stereo localization error by 23 percent with respect to *Mei* and 28 percent with respect to *Scaramuzza*.

We use the accurate non-central geometric model as base model for the proposed centered projection model. For the centered model (7), the non-central geometric model with optimized camera location (3) is selected as base model. The localization performance for the centered model is nearly the same as for the base model, while being significantly faster as shown in Section 3.3.5. Note that a reprojection error after calibration is not denoted, since the centered model uses the calibration result from the non-central base model and does not minimize the reprojection error of the checkerboard corners itself. After remapping the checkerboard observations the reprojection error would be very large, since the checkerboards are very close to the camera which is not the distance for objects in which we are interested later. This effect is analyzed in the following section.

3.3.4. Approximation Results

In this section, we analyze the approximation error of the centered model experimentally. Again, we use the non-central geometric model (3) from our real experiments as ground truth model to create a simulated test set. To demonstrate that the quality of the centered model for objects which are at the distance of interest is sufficient, we evaluate the projection error of the remapped observations for 3D points in different distances. Therefore, we estimate the viewing rays to various image points with different radii to the image center using the non-central ground truth model. On every viewing ray we compute 3D points for different distances. Afterwards, we compute the error between the remapped observations for a 3D point and the estimated observations by applying the centered model to the 3D points directly. This error is the approximation error we achieve by using the centered projection model to non-central systems since the centered model is only perfect for central systems and points at infinity.

In Fig. 3.8 the reprojection error with respect to the distance of the 3D point for various radii, i.e., distances from image center, is illustrated in solid lines. For reference, we also show the reprojection with respect to the original unwarped observations which corresponds to applying a central model without any distortion model to the non-central problem in dashed lines. In (a) the approximation error is illustrated for our real catadioptric system which approximately deviates the single viewpoint condition by 20 mm in axial and 1 mm in lateral direction. Moreover, the approximation errors for a simulated non-central system in lateral direction with 10 mm deviation (c) and in axial direction with 20 mm deviation (d) are shown. In (b) the different radii on the residual field are depicted, with the smallest circle with 100 pixels radius and the largest with a radius of 700 pixels.

The figure shows that the error with the centered model for distances smaller than 100 mm is similar to the error with a central model, but for distances above 1 m the error falls below 0.1 pixel and below 0.01 pixel at 10 meters distance, even though the single viewpoint has been violated by 20 millimeters. Thus, the quality of the centered model degrades enormously and errors introduced by the approximation are very small for the distance of interest.

Further results concerning the centering idea for our real experimental setup are shown in Fig. 3.9. In (a) the viewings rays and the optimal viewpoint for this configuration are depicted, while (b) shows the translation error due to the optimal viewpoint depending on the pixel radius in the image. The residual field for this configuration is illustrated in (c) and the

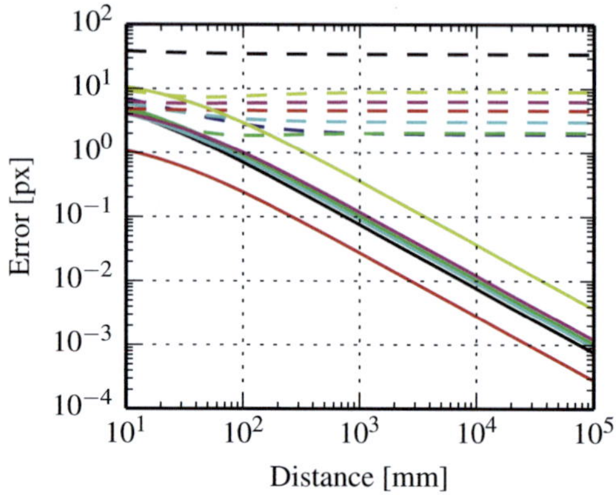

(a) Approximation for our slightly non-central real camera system ($\approx$ 20 mm axial, $\approx$ 1 mm lateral)

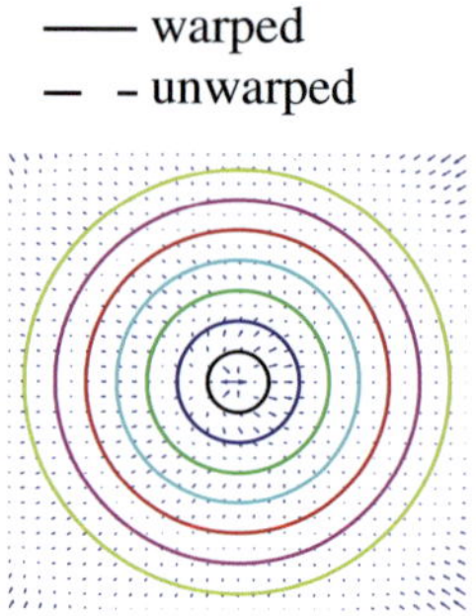

(b) Various Radii on Residual Field

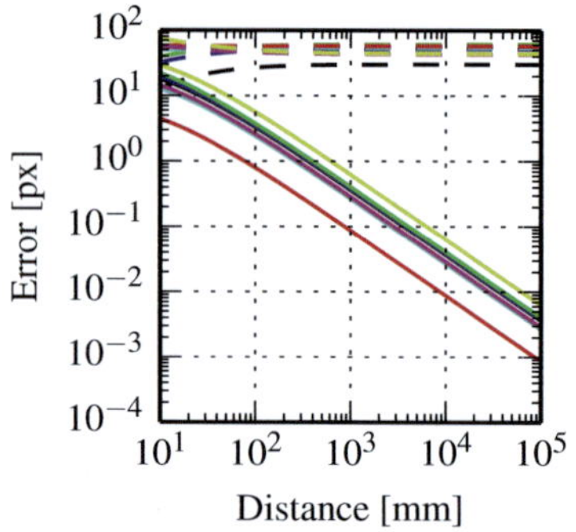

(c) Approximation for simulated deviation in lateral direction (10 mm)

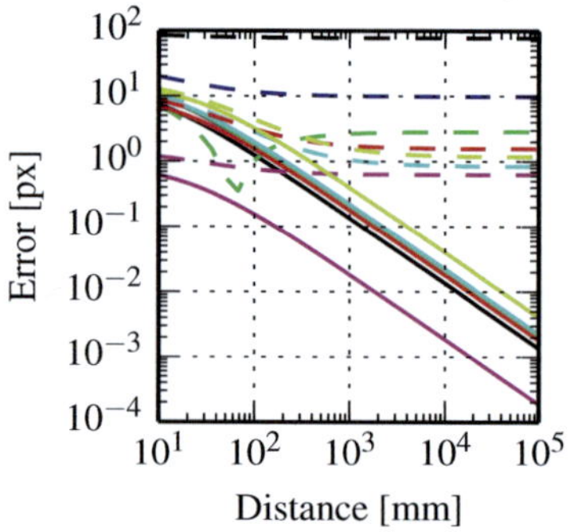

(d) Approximation for simulated deviation in axial direction (20 mm)

Figure 3.8.: **Approximation Error.** This figure shows the average approximation error of the centered projection model in pixels over the distance of the 3D point cloud for our real catadioptric system (a) as well as for a simulated non-central system in lateral (c) and axial direction (d). The corresponding colors to the radii in the image are shown in (b) with the smallest circle radius with 100 pixels and the largest with 700 pixels.

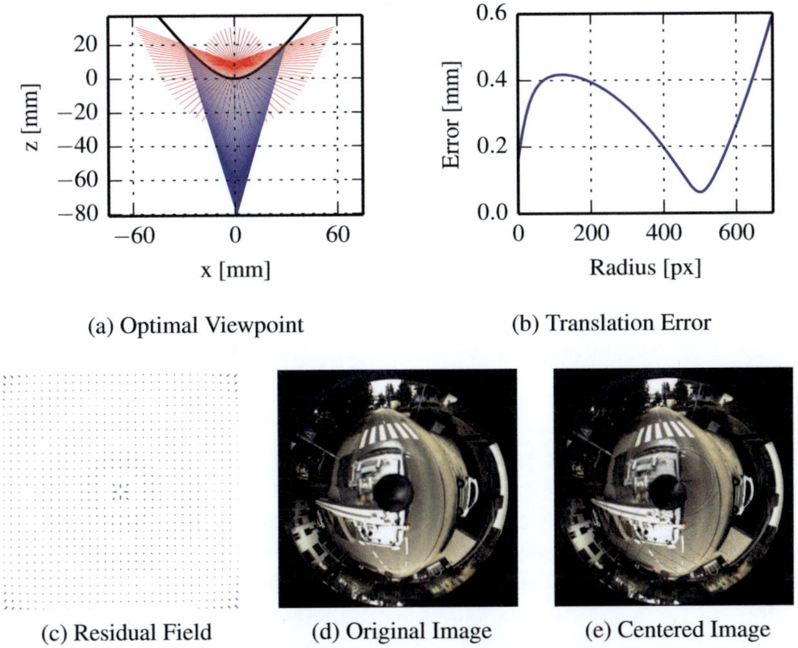

(a) Optimal Viewpoint (b) Translation Error

(c) Residual Field (d) Original Image (e) Centered Image

Figure 3.9.: **Centered Approximation for our Real Camera.** This figure shows the optimal viewpoint (a) for our real experimental setup, the translation error of the viewpoint depending on the radius of the image points (b) as well as the residual field (c) and the original catadioptric image before (d) and after remapping the observations (e). Note, due to the small displacement field the difference between the remapped and the original image is rather small.

residual field applied to the original catadioptric image (d) is shown as the centered catadioptric image in (e). For the deviation in our real experiments which are primarily in axial direction the residuals are very small and the remapped image looks similar to original one. The residuals increase particularly for larger deviation in lateral direction as shown in Fig. 3.2.

3.3.5. Runtimes

In robotics or autonomous applications real-time performance is desirable. Therefore, the projection function which is frequently needed whenever the reprojection error is computed and optimized, e.g., during localization, ego-

	Runtime
Numeric Non-Central [49]	$\sim$ 185,000.00 ms
Geometric Model*	2,919.98 ms
Scaramuzza et al. [96]	913.93 ms
Mei & Rives [75]	6.58 ms
Centered Model	**3.42 ms**
Centered Model*	**1.79 ms**

Table 3.2.: **Runtime.** This table shows the runtimes for projecting 10 000 points in MATLAB with the different projection models. Methods marked with an asterisk (*) are wrapped in C++. The proposed centered model which is as accurate as the geometric non-central model is more than three orders of magnitude faster.

motion estimation, or 3D reconstruction, should be very fast. We analyze the required computation time for the proposed centered projection model and the reference models. For completeness we also present the runtime for a fast numeric non-central model taken from [49] and considered as approximate. To compare the methods, we measure the time to project 10 000 randomly selected 3D points to the image plane. The results are shown in Table 3.2. We compute the runtimes on a standard laptop using a single core with the projection function being MATLAB code. For the geometric model some parts are wrapped in C++ as well as for the centered model marked with an asterisk.

The geometric non-central model is much faster than the numeric non-central model, but it is still very slow. The runtime of the geometric model is heavily dominated by the analytical computation of the reflection point on the mirror surface which involves the computation of the polynomial coefficients and the evaluation of MATLAB's roots function for finding the polynomial roots. The latter one is also responsible for the relatively slow evaluation of the central reference function from Scaramuzza et al. [96]. In contrast, the proposed centered projection function as well as the central projection function from Mei and Rives [75] are very fast and project 10 000 3D points in less than 7 milliseconds only.

Thus, the proposed centered catadioptric projection model is as fast as a central catadioptric projection model and has the accuracy of a non-central model. More precisely, the centered model has nearly the same accuracy as the non-central geometric model, which is used as base model, but speeds-up the projection of more than three orders of magnitude compared to the geometric model.

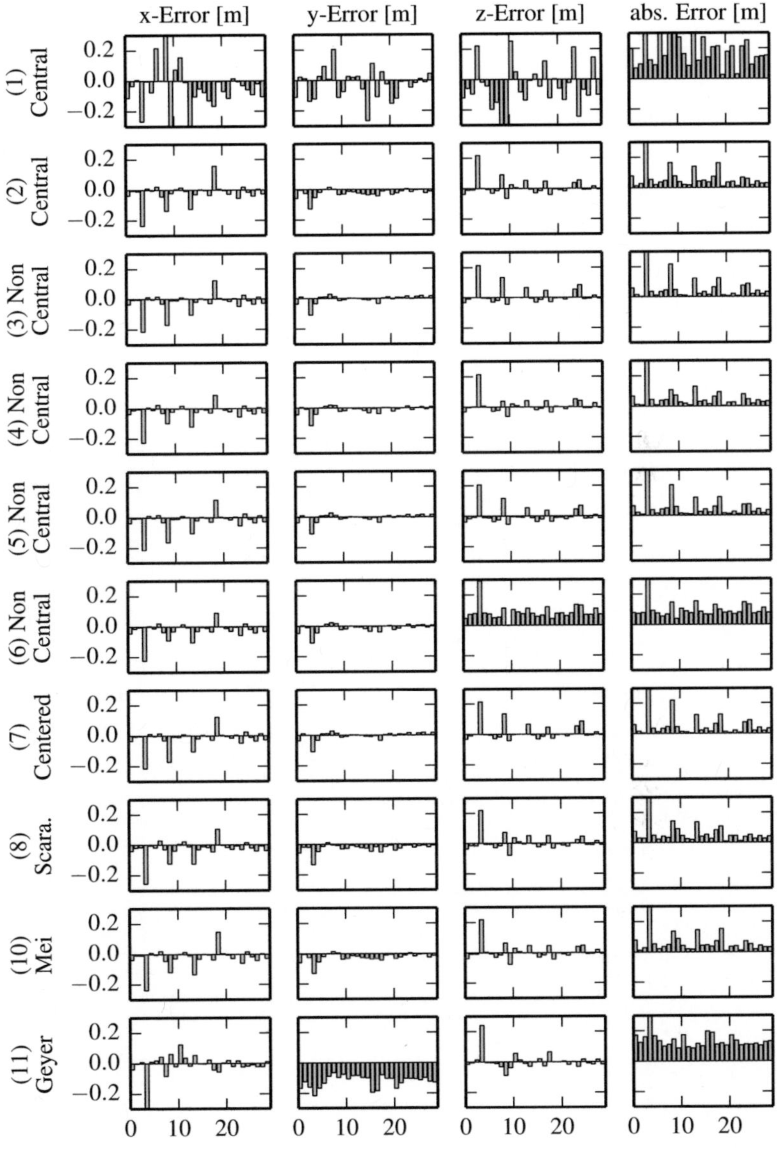

Figure 3.10.: **Mono Localization.** Each diagram shows the error for all 29 (non-collinear) landmark triplets in case of mono localization with the different projection models.

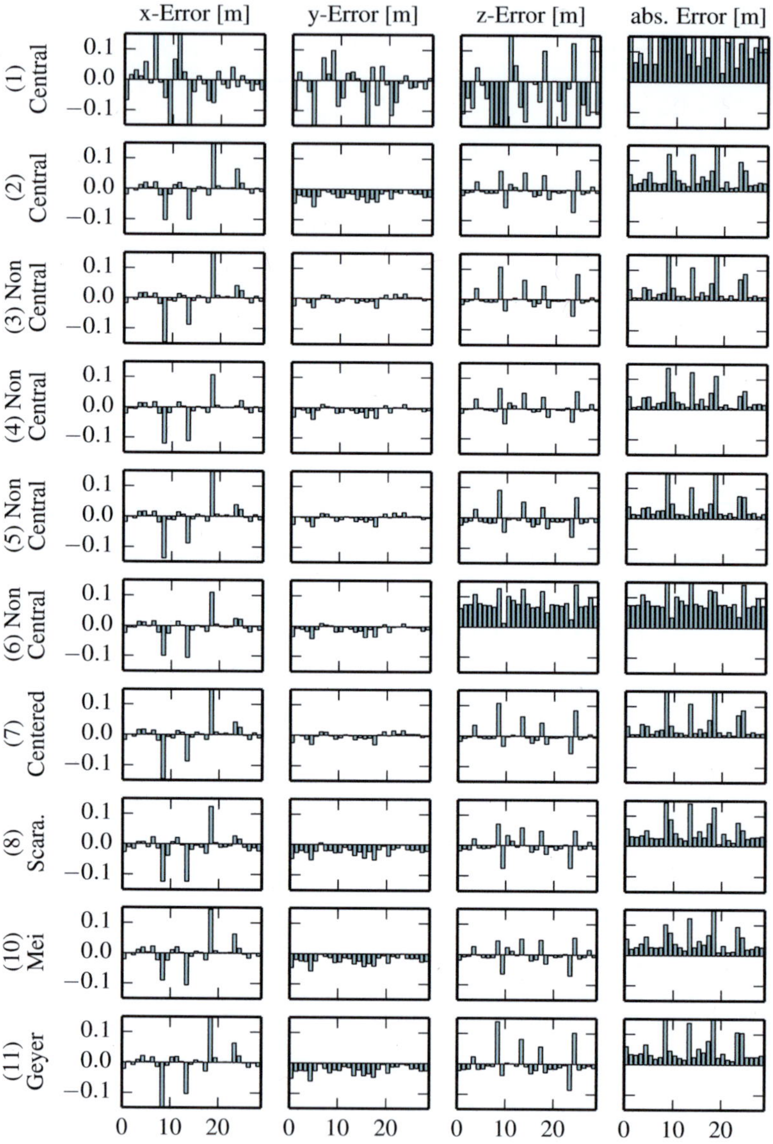

Figure 3.11.: **Stereo Localization.** Each diagram shows the error for all 29 (non-collinear) landmark triplets in case of stereo localization with the different projection models.

Chapter 4

Ego-motion Estimation

In this chapter, we present an ego-motion estimation algorithm for catadioptric stereo cameras which benefits from the proposed novel projection model. Ego-motion estimation is the process to determine the motion of an ego-vehicle between two poses and is a highly relevant application for autonomous vehicles. We show that omnidirectional cameras overcome major drawbacks of traditional perspective cameras for this task. This advantage stems from the greatly extended field of view which we show is crucial for achieving high accuracy motion estimates.

There have been many approaches investigating motion estimation with perspective cameras. However, these approaches suffer from the limited field of view and most approaches capture only the area in front of the ego-vehicle. In this chapter, we show the advantage of ego-motion estimation with catadioptric cameras which provide a 360° field of view. In particular, we use the ego-motion application in outdoor scenarios with real catadioptric cameras which are slightly non-central in most cases, to show the advantages of the proposed centered projection model in contrast to ego-motion estimation with common central catadioptric projection models.

We start this chapter with a brief overview of existing ego-motion estimation methods for omnidirectional cameras. Afterwards, we present the sensor setup for our real-world experiments and explain the proposed motion estimation method for catadioptric cameras. Finally, we conduct a comparative study on different feature matching techniques which is a prerequisite for any motion estimator. We show motion estimation results obtained with catadioptric cameras and the proposed projection model in contrast to results obtained with perspective cameras and with catadioptric cameras with central projection models. The proposed high precision motion estimator allows to accurately stitch top view images computed from catadioptric images. We present some experimental results of several such high fidelity top view maps.

4.1. State-of-the-Art

Perspective cameras, either as monocular or stereoscopic setup, are very popular for visual ego-motion estimation which is also called visual odometry [86]. Moreover, there are many approaches using perspective cameras for the related topic structure from motion [53] where the camera poses as well as the 3D structure is recovered. Simultaneous localization and mapping (SLAM) [27], which is the estimation of the ego-motion while simultaneously updating the map of the surrounding, and high precision ego-vehicle localization [66] has attracted considerable attention lately.

Recently, numerous approaches which propose to use omnidirectional cameras for motion estimation were presented and obtain suitable results. The works include applications for visual odometry [47, 23, 15], structure from motion [21, 70] and SLAM [117, 92]. In [21] the authors demonstrate that omnidirectional cameras outperform perspective cameras for structure from motion particularly in estimating the translation part if the objects are far away. They use a criteria based on the epipolar geometry to estimate the ego-motion. In difference Lhuiller [70] proposes bundle adjustment minimizing an angle and reprojection image error to optimize the ego-motion. A similar approach was used in [78] where the authors minimize the 3D point error for bundle adjustment. Some authors [117, 10] suggest to decouple the estimation of the rotation and translation to increase efficiency and accuracy.

However, most works on omnidirectional cameras use only monoscopic camera setups to estimate the motion from 2D bearing data where scale cannot be estimated from two frames only. There has been little work on stereo omnidirectional motion estimation with catadioptric cameras. In Chapter 1.1 some works using vertical or horizontal stereoscopic catadioptric camera systems are mentioned. However, the approaches using a horizontal stereo setup [40, 37, 29] which is suitable for applications for vehicles are not used to recover motion or 3D structure from the catadioptric image directly.

In general, the existing approaches to estimate motion between two poses can be divided in feature based [23, 70, 95, 117] and appearance based methods [100]. The feature based approaches from Corke et al. [23], Lhuiller [70] and Scaramuzza et al. [95] use a catadioptric camera to capture the whole environment, while Tardif et al. [117] use a *Ladybug* as omnidirectional camera. In this work, we focus on stereoscopic feature based motion estimation. In context with feature based methods for catadioptric cameras there are some approaches [95, 120, 70] which use common local feature detectors, like scale invariant feature transform (SIFT) [72], speeded up robust features (SURF) [9], or Harris corners, and obtain sufficient feature matches. There are also some approaches [99, 30] using simple line features for the feature extraction. Arican and Frossard [4] and Hansen et al. [52] suggest special local feature detectors and descriptors for catadioptric images mostly based on the classical feature detectors and descriptors.

There are several methods to improve the ego-motion estimation result more or less independently from the different estimation approaches. Motion constraints can be used to decrease computational time and improve the accuracy. In [95] the authors assume a planar and circular motion to parameterize the motion with only one feature correspondence. Moreover, outliers can be removed with the random sampling consensus (RANSAC) algorithm [32] which has been established as an iteratively standard method to obtain model parameters from data with outliers. Since the motion is estimated incrementally, the errors are accumulated over time which introduces drift of the estimated trajectory compared to the real trajectory. This drift can be reduced by bundle adjustment [118], the local optimization over the last frames, or by loop closure [24], the detection of previously visited places.

However, in this work we focus on the two-frame unconstraint motion estimation. We refrain from using any bundle adjustment or loop closure detection as our primary goal is to demonstrate the feasibility of using a catadioptric stereo camera setup with our novel proposed centered projection model.

4.2. Sensor Setup

For the applications, more precisely the ego-motion estimation presented in this chapter and the dense 3D reconstruction presented in the next chapter, we use a similar sensor setup as for the calibration evaluation (Chapter 3.3.1). In particular, we use the same hypercatadioptric cameras where the camera position deviates from the optimal position approximately by 20 mm in axial direction. We mounted two of these cameras on top of our driving platform AnnieWAY such that they are horizontally aligned as shown in Fig. 4.1a. The baseline between the cameras is approximately 0.8 meters.

(a) Driving Platform AnnieWAY

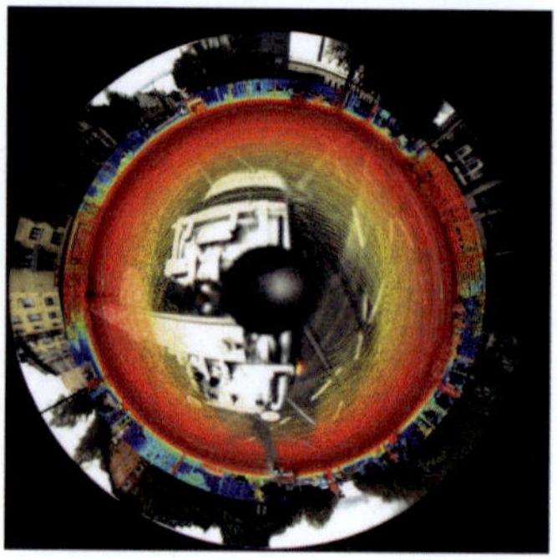

(b) Projected Velodyne Points

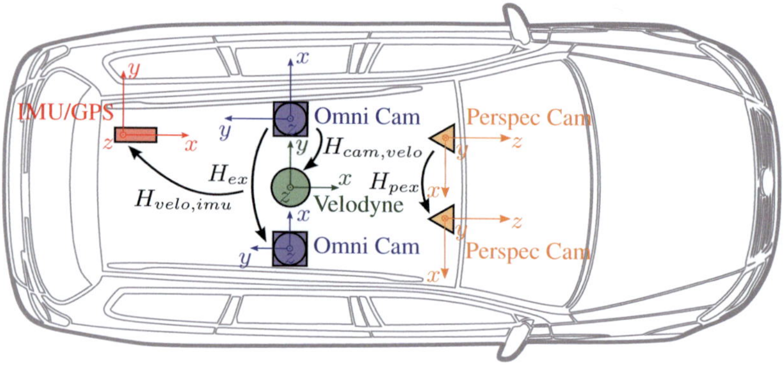

(c) Sensor Setup

Figure 4.1.: **Recording Platform.** This figure (a) shows our driving platform AnnieWAY with two horizontally aligned hypercatadioptric cameras. In (b) the reprojected Velodyne point cloud to the catadioptric image is shown, where the color denotes the depth of the point. A top view of the sensor setup with the corresponding coordinate systems as well as the transformations between the sensors is depicted in (c).

Furthermore, the platform is equipped with a high precision GPS/IMU system that delivers ground truth motion and a *Velodyne HDL-64E* rotating 3D laser scanner that provides laser scans with a horizontal field of view of $360°$ and a vertical resolution of 64 laser beams. Thus, the laser scanner delivers a 3D point cloud of the environment. We use the 3D point cloud of the laser scanner as accurate ground truth 3D information.

Moreover, the vehicle contains a perspective stereo camera system facing in frontal direction of the ego-vehicle. The perspective cameras are *Flea2* 1.4 Megapixels color cameras and have a baseline of approximately 0.6 meters. All cameras are triggered by the Velodyne when it is facing forward. This induces a frame rate of 10 Hz (fps) for the cameras. The IMU has a frame rate of 100 Hz and the closest time stamp is chosen for synchronization. With these setup of two catadioptric and two perspective cameras, a laser scanner and the IMU/GPS system we captured different urban scenarios.

The different sensors need to be intrinsically and extrinsically calibrated with respect to the reference coordinate system which we have chosen to be the left catadioptric camera coordinate system. The transformation between the sensors as well as the coordinate systems of the different sensors are shown in Fig. 4.1c. The catadioptric camera calibration is achieved with the presented catadioptric stereo calibration toolbox using planar checkerboards in different positions and orientations. The extrinsic calibration (rotation and translation) between the two catadioptric cameras is denoted as H_{ex}.

For the extrinsic calibration $H_{cam,velo}$ of the Velodyne with respect to the catadioptric reference camera, we reproject the laser scanner point cloud to the corresponding catadioptric image with a manually chosen initial transformation. We obtain the exact transformation by a non-linear optimization with the Levenberg-Marquardt algorithm similar to the optimization of the catadioptric camera parameters. We minimize the Euclidean error between the catadioptric image points and the reprojected image points from the 3D laser points of 50 manually selected correspondence points. For the correspondence points we choose corners which are visible in the reprojected point cloud from the laser scanner and in the catadioptric image. The projected laser point cloud in the catadioptric camera image is shown in Fig. 4.1b, where the color of the projected point denotes the depth. The transformation between the GPS/IMU and the Velodyne $H_{velo,imu}$ as well as the calibration of the perspective cameras H_{pex} are obtained with the approaches in [41, 42].

4.3. Ego-motion Estimation

In this work, we present a stereo ego-motion estimation method with two catadioptric cameras which estimates the motion between the poses of two consecutive frames. We refrain from using any sophisticated method such as bundle adjustment, loop closure, or a motion model and focus on two-frame motion only to demonstrate the feasibility of using catadioptric cam-

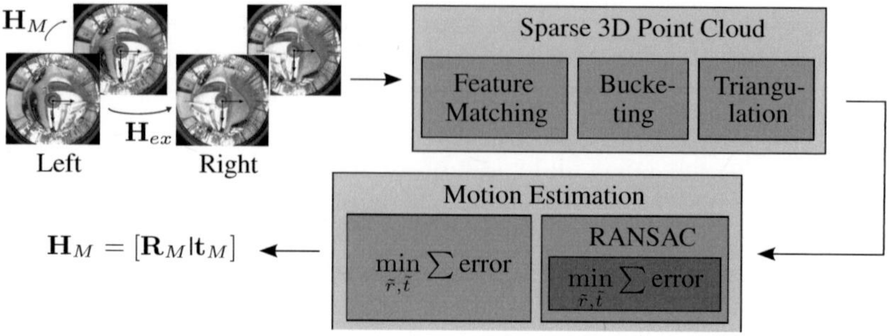

Figure 4.2.: **Ego-motion Estimation.** This figure shows the ego-motion estimation method divided into the two steps: Sparse 3D point cloud estimation and the motion estimation process. The input of the method are two consecutive stereo image pairs and the intrinsic and extrinsic calibration of the stereo camera. The result of the method is a six DOF motion vector.

eras with the proposed projection model. However, we expect that improvements in the simple task directly translate into improvements when using more advanced algorithms. In particular, loop closure detection in catadioptric images, where also loops in the opposite driving direction could be detected, reduces the drift of the resulting trajectories.

An overview of the presented ego-motion estimation method is shown in Fig. 4.2. We use two consecutive stereo image pairs (for time step $t-1$ and t) as well as the proposed intrinsic and extrinsic catadioptric calibration result as input for the method. The ego-motion estimation method with a stereoscopic catadioptric camera system is divided into two main parts. In a first step a sparse 3D point cloud from the previous frame $(t-1)$ is estimated and in the second part we calculate the ego-motion between the poses in the current (t) and previous frame. Finally, we achieve a six degree of freedom (DOF) motion vector not constrained to planar motion. Thus, we estimate a 3D translation vector $\mathbf{t}_M$ in x-, y- and z-direction as well as 3D rotation vector $\mathbf{r}_M$ which represents a rotation matrix in the Rodrigues formulation. In the following the two main parts are explained in detail.

4.3.1. Sparse 3D Point Cloud

In the first part, we compute a sparse 3D point cloud from two consecutive stereo image pairs. Therefore, corresponding features in all four catadioptric images are matched. A similar matching strategy to the *circle* matches for perspective images in [43] is applied. Hence, we initially find a match between the previous left and previous right image. Then, we search for the corresponding match for the selected feature point in the previous right image in the current right image, then for the corresponding matches between the current right and current left image and finally for the match between the current left and previous left image. A feature match is only taken as a valid candidate, if the feature point in the previous left image from the last matching step coincide with the feature point from the first matching step.

We concentrate on local point features for the correspondence search between the images which allow the computation of a 3D position. In contrast, line features allow the computation of a 2D position without the height of the world point only. For the feature detection and matching we use common matching strategies which achieve good performance in perspective image pairs [79, 80]. Some works [95, 120] partly use this features in catadioptric images, however, an evaluation concerning the performance of this features in catadioptric images against ground truth is still missing. We fill this gap and conduct a comparative study on feature matching on catadioptric images using high precision ground truth. We evaluate the performance of different common feature detector and descriptor combinations on catadioptric images.

For the detection of interesting feature points, we use the popular SIFT (Scale Invariant Feature Transform) [72] and the similar but faster SURF (Speeded Up Robust Features) [9] detector. We also use the very efficient FAST (Features from Accelerated Segment Test) [93] feature detector and his extensions ORB (oriented FAST and rotated BRIEF) [94], which also computes the orientation, and BRISK (Binary Robust Invariant Scalable Keypoints) [68], which search maximas in the 3D scale space. To describe the feature points, we use the rotation and scale invariant common SIFT and SURF descriptors which are unfortunately too slow for real-time applications due to the expensive calculation of gradients. We also evaluate the recently proposed binary descriptors BRIEF (Binary Robust Independent Elementary Features) [19], ORB, and BRISK, which are very fast to compute and match, since they use simple binary tests between pixels and the Hamming distance instead of the L^2-norm as a distance measure value. In Chapter 4.4.1 we show the performance evaluation of the different feature matching strategies. For the evaluation of our motion estimation method we use an appropriately performing feature detector.

As we do not have any prior knowledge about the movement in temporal direction, we cannot use the epipolar geometry to simplify the correspondence search. Consequently, we search feature matches in the whole catadioptric image, accept for the black margin and the ego-vehicle itself. Since we do not use bundle adjustment over a larger number of frames and observe only feature matches between two neighboring frames, we reduce

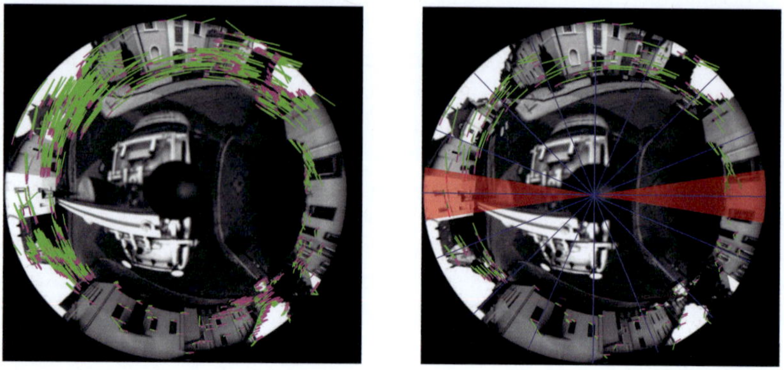

(a) Matches before Bucketing

(b) Matches after Bucketing

Figure 4.3.: **Feature Matching.** In (a) the catadioptric image with corresponding feature matches from all four images is depicted. The spatial stereo correspondences are shown in green and the temporal correspondences in magenta. In (b) the remaining feature matches after bucketing are shown as well as the cells for the bucketing algorithm in blue. Feature matches in the red area are ignored due to the bad reconstruction accuracy in this area.

the outliers (false matches) by constraining the movement of corresponding points in the images with a maximum distance between matching features.

Moreover, we use a bucketing technique to reduce the number of correspondences and achieve a uniform distribution of the feature points in the image. The reduced feature number speeds up the ego-motion estimation while a better distribution improves the result of the ego-motion avoiding biases. In practice, we divide the catadioptric image in 16 cells depending on the azimuth angle shown in Fig. 4.3b. For each cell we allow a maximum of 12 feature matches. Consequently, we have a maximum number of 192 matches for further processing. In Fig. 4.3 the result of the correspondence search in spatial and temporal direction before (a) and after (b) bucketing is depicted. The spatial stereo matches are shown in green and the temporal matches in magenta.

Afterwards, we estimate the initial 3D world points from the 2D image correspondences from the previous left and right image with triangulation. A 3D point can be computed as the midpoint of the shortest distance between the left and right reflecting viewing rays. Thus, we compute for each image point the reflecting ray on the mirror surface going through the optimal single viewpoint. To compute the reflection rays we use the inverse central-centered projection function $Q^{-1}(u, v)$ (see Eq. 2.34).

Reconstruction Accuracy The reconstruction accuracy of the triangulation from two horizontally aligned catadioptric cameras depends on the 3D world point position of the estimated point, particularly on the azimuth angle φ of the viewing rays. We have presented a detailed analysis for this dependency for two [104] and three [105] cameras. To show this dependency, we simulate a 3D environment and perform a Monte Carlo simulation. Herein, we simulate a circular point cloud with constant height around the midpoint of the camera baseline and project the 3D points to the image plane. Afterwards, we create 100 samples of each image point and disturb each sample with Gaussian noise of standard deviation 1.5 pixels. Then, we perform the triangulation of each sample to achieve the 3D point and subsequently reconstruct each 3D point to the image plane. We interpret the accuracy by computing mean and standard deviation of the Euclidean error between the reconstructed and the ground truth image point.

Fig. 4.4b shows the normalized mean Euclidean error for a circular point cloud with radius eight meters around the cameras and constant height depending on the azimuth angle. The distribution of the 3D reconstruction for all samples of the point cloud is depicted in Fig. 4.4a. The error is maximal at $\varphi = 90°$ and $\varphi = 270°$ respectively, which coincide with the baseline between the cameras. Subsequently, a reconstruction along the baseline between the two cameras is not possible as expected. The simulation results allow us to make a judicious selection of the area used for feature matches. In Fig. 4.3b we mark this area where we do not use the feature matches in red.

4.3.2. Motion Estimation

After estimating the sparse 3D point cloud by triangulation in the previous frame, we compute the ego-motion of the vehicle between the current (t) and previous $(t-1)$ frame by minimizing the reprojection error in the image. As already mentioned, we consider six motion parameters, a 3D translation vector $\mathbf{t}_M$ and a 3D rotation vector $\mathbf{r}_M$ resulting in a rotation matrix $\mathbf{R}_M$. We transfer the triangulated 3D points $\mathbf{p}_{l,t-1}$ from the previous reference frame in the current left frame $\mathbf{p}_{l,t}$ and in the current right frame $\mathbf{p}_{r,t}$ with

$$\begin{bmatrix} \mathbf{p}_{l,t} \\ 1 \end{bmatrix} = \mathbf{H}_M \begin{bmatrix} \mathbf{p}_{l,t-1} \\ 1 \end{bmatrix} \tag{4.1}$$

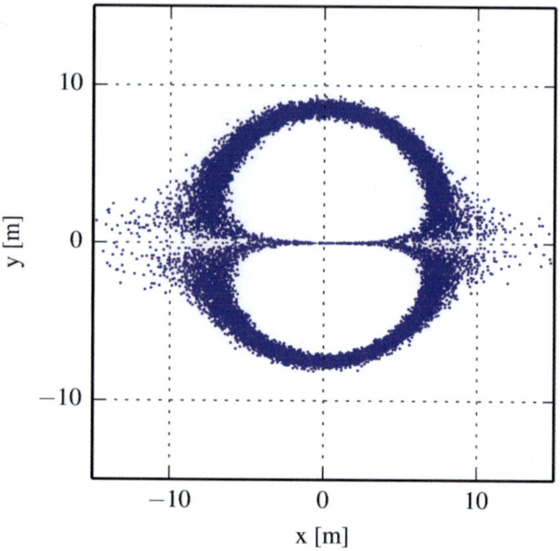

(a) Simulated reconstruted circular point cloud

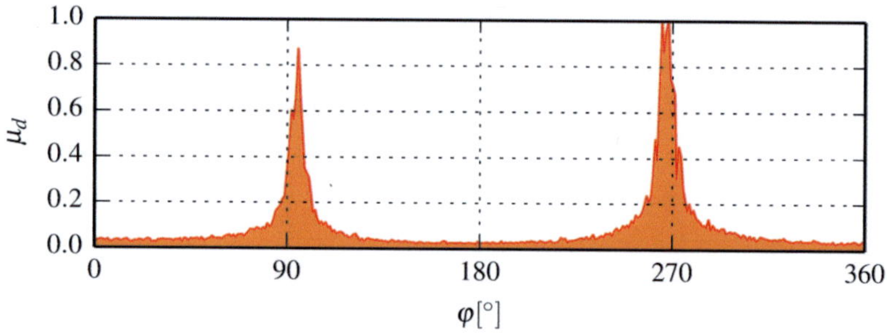

(b) Normalized mean Euclidean error μ_d for different azimuth angles φ

Figure 4.4.: **Reconstruction Accuracy.** This figure shows the dependency of the reconstruction accuracy from the azimuth angle φ of the viewing ray. In (a) the reconstruction result for all samples of a simulated noisy circular point cloud with 8 m radius and constant height around the cameras is shown. (b) depicts the normalized mean Euclidean error μ_d as a function of the azimuth angle φ .

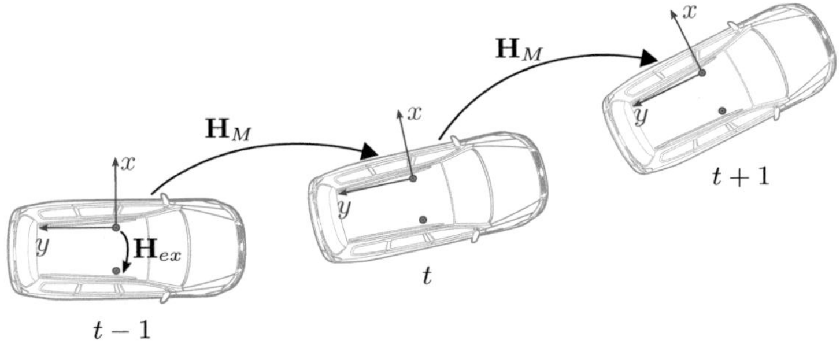

Figure 4.5.: **Ego-motion Estimation.** This figure shows the rigid motion between the ego-vehicle at three different time steps where $\mathbf{H}_M$ is the motion between two consecutive frames.

$$\begin{bmatrix} \mathbf{p}_{r,t} \\ 1 \end{bmatrix} = \mathbf{H}_{ex}\mathbf{H}_M \begin{bmatrix} \mathbf{p}_{l,t-1} \\ 1 \end{bmatrix} \tag{4.2}$$

where

$$\mathbf{H}_M = \begin{bmatrix} \mathbf{R}_M(\mathbf{r}_M) & \mathbf{t}_M \\ \mathbf{0} & 1 \end{bmatrix} \tag{4.3}$$

is the rigid motion between two frames in homogeneous coordinates and $\mathbf{H}_{ex}$ the transformation between the left and right camera. The relationship between an ego-vehicle at different time steps is illustrated in Fig. 4.5. The motion parameters $\mathbf{r}_M$ and $\mathbf{t}_M$ between two frames are obtained by a non-linear optimization using the Levenberg-Marquardt algorithm. We iteratively minimize the reprojection error for all feature correspondences n_{cor} yielding

$$\mathbf{r}_M, \mathbf{t}_M = \underset{\tilde{\mathbf{r}}_M, \tilde{\mathbf{t}}_M}{\operatorname{argmin}} \left\{ \sum_{i=1}^{n_{cor}} \underbrace{\left\| \mathbf{q}_{l,i}^{(E)} - \mathbf{q}_{l,i}^{(*)}(\mathbf{p}_{l,t}(\mathbf{r}_M, \mathbf{t}_M)) \right\|_2^2}_{\text{left camera}} + \right.$$
$$\left. \lambda \underbrace{\left\| \mathbf{q}_{r,i}^{(E)} - \mathbf{q}_{r,i}^{(*)}(\mathbf{p}_{r,t}(\mathbf{r}_M, \mathbf{t}_M)) \right\|_2^2}_{\text{right camera}} \right\} \tag{4.4}$$

with $\lambda \in \{0, 1\}$. The matched feature points in the left image are denoted by $\mathbf{q}_l^{(E)}$, while the projected image points in the left image are $\mathbf{q}_l^{(*)}$. The matched and projected points in the right image are denoted as $\mathbf{q}_r^{(E)}$ and $\mathbf{q}_r^{(*)}$ respectively. The projected points are obtained by projecting the estimated 3D points depending on the motion parameters with the centered projection model (see Eq. 2.33) or the central reference models.

In our experiments we consider two cases: We perform the motion estimation after triangulation from both cameras in the previous frame by minimizing the reprojection error in the current left reference camera image ($\lambda = 0$) only and by minimizing in both current camera images ($\lambda = 1$). For both cases we initialize the motion parameters $\mathbf{r}_M$ and $\mathbf{t}_M$ to zero which is sufficient for convergence after few iterations. To achieve robustness against outliers, we use the motion estimation method within a RANSAC scheme. Thus, we first estimate the motion parameters for m iterations independently using three randomly selected correspondences, since this is the minimum number of feature correspondences to estimate the six motion parameters. In our experiments we use 50 iterations which is sufficient to obtain an outlier free subset with adequate probability. For each RANSAC iteration we compute the reprojected feature points which have an error smaller than a threshold of five pixels and count these points as inliers. Afterwards, we use all inliers of the iteration with the maximum number of inliers and estimate the motion parameters with a final non-linear optimization step.

4.4. Evaluation

We evaluate the presented motion estimation method on different urban scenarios captured with the presented sensor setup. In particular, we show the benefit of the centered projection model compared to common central projection models with this application.

First of all, we evaluate the feature matching on real catadioptric images since an evaluation of feature matching strategies on catadioptric images is missing in the literature. Afterwards, we use an appropriate feature matching method to evaluate the ego-motion estimation for the centered projection model in comparison to popular state-of-the-art central catadioptric projection models. For a quantitative result we use the GPS/IMU trajectories as ground truth. Moreover, we compare the ego-motion results from a stereoscopic catadioptric camera setup against the results of a perspective stereo camera setup capturing the same scene. In the end, we show the resulting accumulated trajectories on high fidelity top view maps created from the catadioptric images.

Detector	Number	Runtime [ms]
SIFT	2118	1646.60
SURF	**5058**	1145.53
FAST	3081	**12.11**
ORB	423	83.62
BRISK	1074	132.70

Table 4.1.: **Feature Detector Evaluation.** This table shows the mean number of detected keypoints and the mean runtime for different feature detectors evaluated on 1 000 randomly selected image pairs from different urban sequences.

4.4.1. Feature Matching

Before we evaluate the ego-motion estimation, we analyze which standard feature matching strategies for perspective images can also handle the large image distortions and blur in catadioptric images. Therefore, we evaluate different feature detectors and descriptors which have shown good performance for standard perspective cameras [80, 79]. For the different detector and descriptor methods we use the OpenCV implementation [13]. Since two consecutive frames are used for the ego-motion estimation, we have a small baseline between two temporal frames similar to the baseline in a spatial image pair. Thus, we only evaluate the matching strategies between two spatial neighboring catadioptric image frames.

The performance of the different matching strategies is evaluated with the Velodyne laser scanner. We reproject the 3D Velodyne point cloud to both images of the catadioptric image pair. To achieve a dense 3D ground truth point cloud, we accumulate the laser point clouds of seven frames with an ICP point-to-plane fitting [41]. This constrains the evaluation to static scenes but increases the probability to find a corresponding ground truth point to the matched feature point. The vertical field of view of the Velodyne is much smaller than the one of the catadioptric camera as shown in Fig. 4.1b. Therefore, we evaluate only feature matches in the overlapping area. For every matched feature point in the first catadioptric image, we estimate the corresponding projected Velodyne point in this image and compute the corresponding ground truth feature point in the second image. We calculate the Euclidean distance in the second image between the computed projected Velodyne correspondence point and the matched feature point as an error of the feature matching.

The detectors and descriptors are evaluated on 1 000 randomly selected image pairs of different urban sequences. In Table 4.1 we show the mean number of detected feature points as well as the mean runtime for the fea-

Detector	Descriptor	Precision	Time [s]	Correct matches	All matches
FAST	SIFT	**0.90**	2.33	205	226
FAST	SURF	0.12	1.21	136	1152
FAST	BRIEF	0.79	**0.74**	279	354
FAST	ORB	0.73	0.75	168	230
FAST	BRISK	0.77	1.63	193	250

Table 4.2.: **Descriptor Evaluation.** This table depicts performance and runtime of different descriptors for the same number of FAST feature points. The values are mean values over 1 000 image pairs from different urban sequences.

ture detection in one image pair for different detectors. The largest number of detected feature points is achieved with the SURF feature detector. However, this detector needs around 1.1 seconds to detect feature points in both images of the catadioptric image pair which is not applicable for real-time applications. The most efficient detector is the FAST detector which also achieves a sufficient number of matches in around 0.01 seconds per image pair.

For a fair comparison of different feature descriptors, we use the same number of keypoints, all extracted by the FAST detector. We evaluate the precision which is computed by

$$\text{precision} = \frac{\text{correct matches}}{\text{all matches}} \tag{4.5}$$

and the mean runtime of the descriptors. As correct matches we count the feature matches which have a smaller error than a threshold of five pixels. Table 4.2 shows the precision and the runtime as well as the mean number of all feature matches and the mean number of correct matches.

Concerning the feature matching, we use the distance measure as proposed for the particular descriptor in the OpenCV implementation and take only the best matches depending on a threshold. For the combination of FAST features and SURF descriptor this threshold based selection does not work well, so the precision is very small in this case. However, although the number of matches is very large only the smallest number of correct matches is found. The best precision is achieved with the SIFT descriptor followed by the BRIEF descriptor. A high precision is desirable, since it reduces the number of required RANSAC iterations in the ego-motion method. Since the BRIEF descriptor is about four times faster than SIFT, we use the BRIEF descriptor for the ego-motion estimation. Due to the small baseline of the image pairs and the involved small deformations in the image, a non-rotation and scale invariant descriptor is sufficient to achieve good feature matches in the catadioptric image pairs. Therefore, we use the FAST corner detector in combination with the BRIEF descriptor for the ego-motion estimation method with catadioptric cameras.

4.4.2. Motion Results for different Projection Models

We evaluate the results for the ego-motion estimation on three urban sequences with different frame lengths in the range of 2 300 to 4 000 frames against the ground truth GPS/IMU motion $\mathbf{H}_{imu}$. Therefore, we transform the ground truth motion in the reference camera coordinate system

$$\mathbf{H}_{gt} = \mathbf{H}_{cam,velo} \cdot \mathbf{H}_{velo,imu} \cdot \mathbf{H}_{imu} \cdot \mathbf{H}^{-1}_{velo,imu} \cdot \mathbf{H}^{-1}_{cam,velo} \quad (4.6)$$

with the calibrated transformation between IMU and Velodyne laser scanner $\mathbf{H}_{velo,imu}$ and the transformation between Velodyne and the left reference camera $\mathbf{H}_{cam,velo}$.

We compare the ego-motion estimation results obtained with the presented centered projection model against the one obtained with the common central models from Scaramuzza et al. [96] and Mei and Rives [75]. For the *Scaramuzza* model we use the improved calibration result instead of the original one as presented in Section 3.2.4. For each projection model the same feature matches combining FAST corner detector and BRIEF feature descriptor are used to compute the motion estimation. We obtain around 1 300 feature matches with the *circle* matching strategy before bucketing is used, depending on the scene information.

For comparing the different projection models, we compute the ego-motion after triangulation with both previous cameras by minimizing the reprojection error in the left reference camera image only (similar to minimizing in monoscopic motion estimation) and in both camera images. The influence of using a central model for slightly non-central cameras is particularly pronounced when minimizing the ego-motion in one camera only. To evaluate the motion estimation we compute the end-point error $\mathbf{e}_i(j) \in \mathbb{R}_3$ of the motion estimation for frame i starting at frame $i - j$. For minimizing in one camera image we choose $j = 200$ frames which approximately corresponds to a driven path of 200 meters depending on the driving speed. In Table 4.3 (left side, rows 1 - 4) we show the end-point errors after 200 frames for minimizing in one camera image only. We compute the mean Euclidean norm end-point error $\|\mathbf{e}_i\|$ for all frames of all sequences as well as the mean end-point errors e_{i_x} in x-, e_{i_y} in y- and e_{i_z} in z-direction. In Fig. 4.6 we illustrate the end-point errors for the three evaluated sequences

Error [m]		One Camera Centered	One Camera Mei	One Camera Scara.	Two Cameras Centered	Two Cameras Mei	Two Cameras Scara.
$j = 200$	$\|\|\mathbf{e}_i\|\|$	**8.84**	14.51	16.53			
	e_{i_x}	2.16	**1.63**	2.07			
	e_{i_y}	**2.40**	2.48	3.19			
	e_{i_z}	**7.72**	14.01	15.89			
$j = 1\,000$	$\|\|\mathbf{e}_i\|\|$	**29.51**	49.38	61.08	**10.61**	11.72	16.90
	e_{i_x}	**8.54**	12.91	22.26	7.16	**6.20**	11.43
	e_{i_y}	**8.05**	11.90	19.81	5.55	**4.94**	8.96
	e_{i_z}	**24.69**	44.04	48.36	**2.57**	6.04	3.52

Table 4.3.: **Motion Estimation Results for Catadioptric Projection Models.** This table shows the end-point error $\mathbf{e}_i(j)$ for an accumulated trajectory estimated by ego-motion estimation with minimizing in only one image (left) or in both images (right) with the different catadioptric projection models. The end-point error is computed for $j = 200$ frames (rows 1 - 4) when minimizing in one image and for $j = 1\,000$ frames (rows 5 - 8) when minimizing in both images. The errors are mean values over all 10 300 frames from all sequences.

for all methods. In Table 4.3 and Fig. 4.6 it is clearly visible that the end-point error of the trajectories computed with the proposed centered model is much smaller compared to the estimation with the central reference models. Mainly the error in z-direction which corresponds to the altitude, is about two times smaller while the error in x- and y-direction is similar.

The same results can be seen in Fig. 4.7 for the accumulated trajectory of the first sequence compared to the ground truth GPS trajectory (black). In (a) a top view of the sequence with 2 300 frames, overlaid on a BING satellite image, is depicted. A side view of a part (the last 1 000 frames) of the same trajectory is shown in (b). While the performance of the motion estimation with all different projection models looks similar in the top view, it is clearly visible that the proposed centered method (blue) performs much better for estimating the z-component of the trajectory. Overall, due to the axial deviation from the single viewpoint condition of approximately 20 mm, the centered model is able to significantly reduce drift in z-coordinate direction for the ego-motion estimation method by minimizing in one camera.

In the same way, we compare the results for the motion estimation minimizing the reprojection error in both images. As expected the errors between the ground truth and the estimated trajectories with all projection models are much smaller than for minimizing in only one camera image

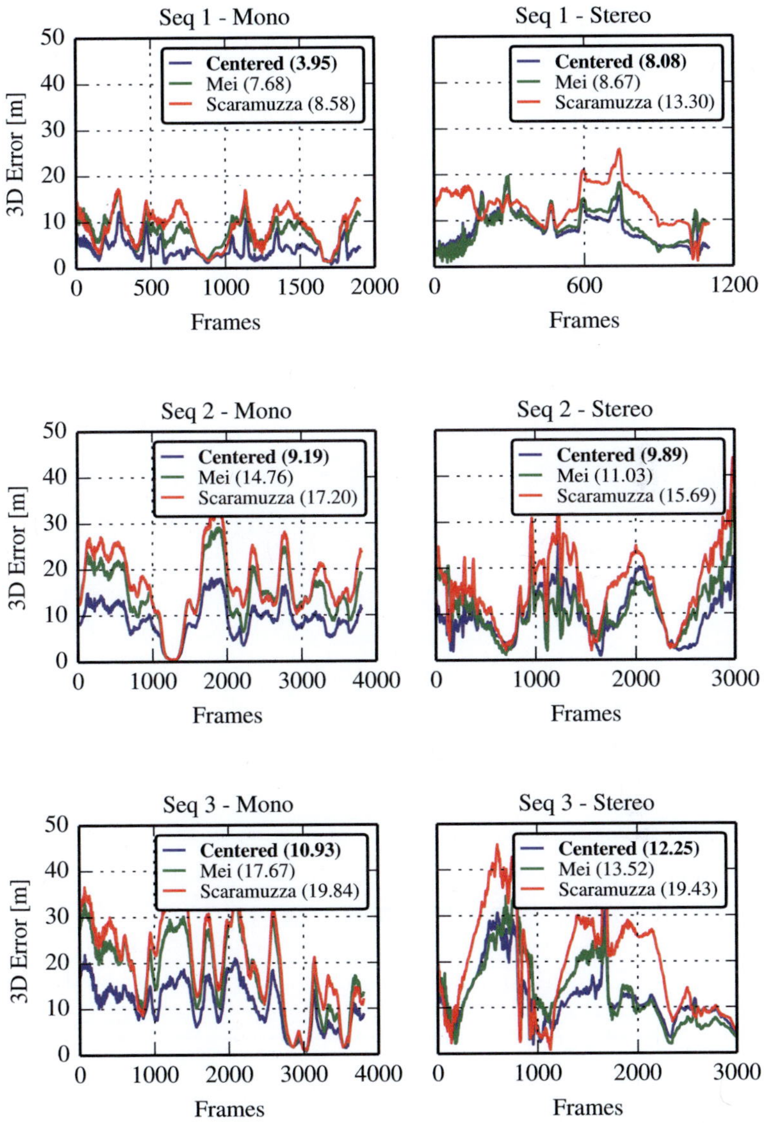

Figure 4.6.: **End-point Error.** This figure shows the end-point errors $\mathbf{e}_i(j)$ for the ego-motion estimation with one camera (left side) after $j = 200$ frames and with two cameras (right side) after $j = 1\,000$ frames for three different urban sequences of varied length.

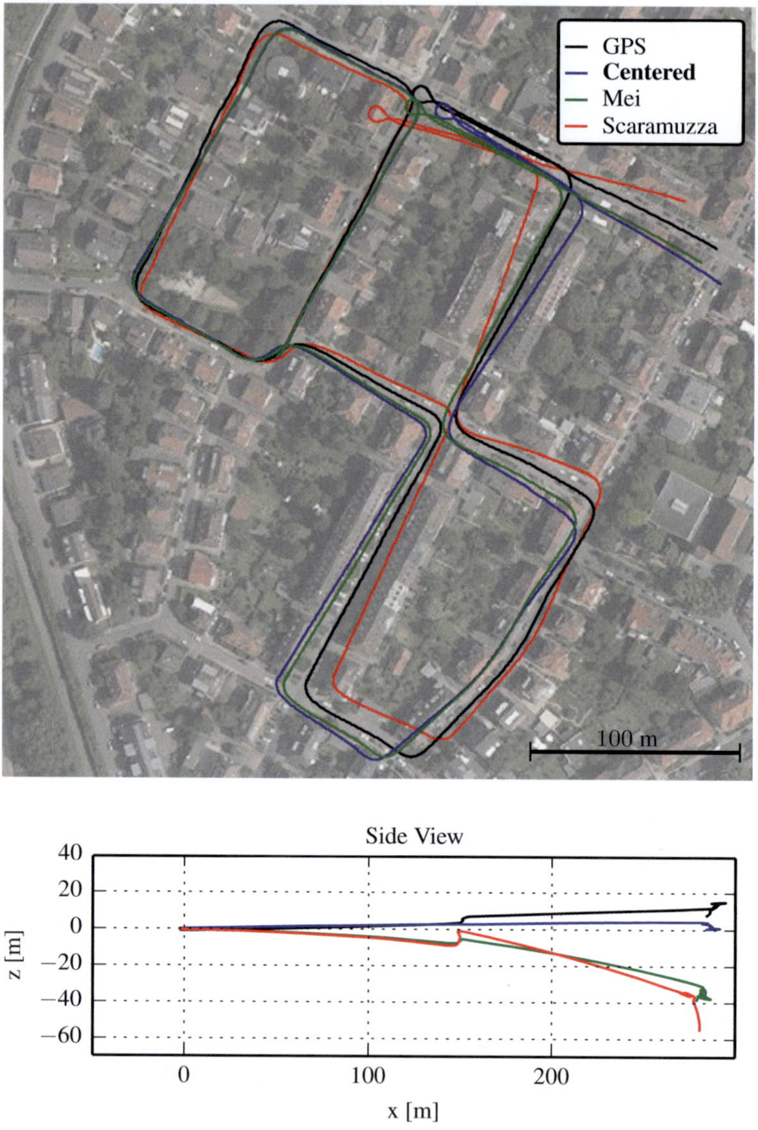

Figure 4.7.: **Ego-motion Estimation One Camera.** This figure shows the estimated accumulated trajectory by minimizing the reprojection error in only one camera with the proposed centered projection model and with the central reference models. The top view of the trajectory (a) shows the whole sequence with 2 300 frames, while the side view (b) shows only a part (last 1 000 frames) of the sequence.

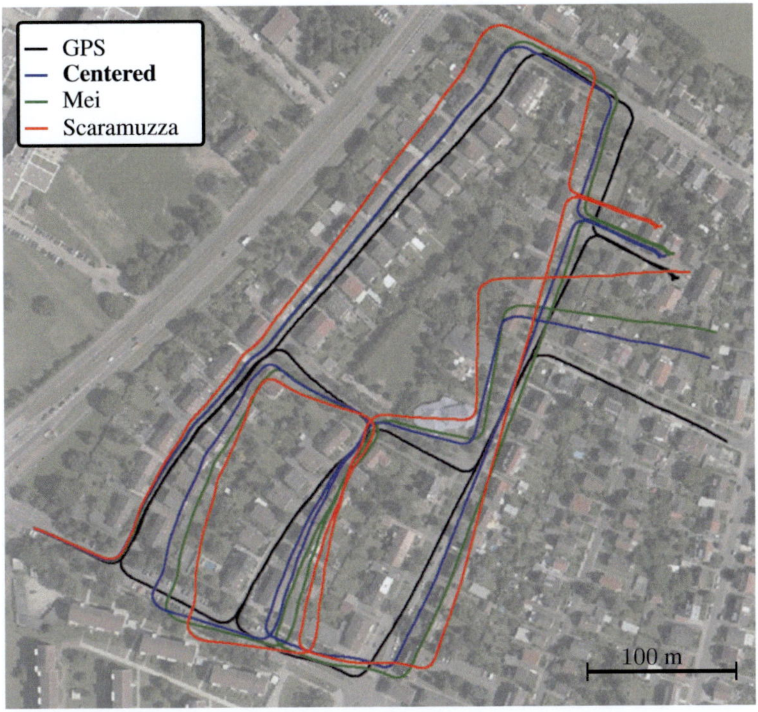

Figure 4.8.: **Ego-motion Estimation Both Cameras.** This figure shows the estimated trajectory of the second sequence (4 000 frames) by minimizing the reprojection error in both camera images with the centered projection model and the central reference models.

as shown in Table 4.3 (rows 5 - 8). Thus, the errors between the ground truth and the estimated trajectory are mainly effected by the quality of the ground truth depending on calibration errors between the sensors. To optimize the calibration, we take the first 100 frames of a separate sequence and optimize the rotation parameters of the transformation between the IMU and the reference camera by minimizing the reprojection error between all poses. For a fair comparison, we do this calibration refinement for each projection method separately and compare the resulting trajectories against the corresponding ground truth. However, this calibration refinement also compensates for possibly existing bias in the motion estimation method itself. Nevertheless, for long sequences we see the same effect, that the drift in z-direction and the mean end-point error are smaller with the proposed

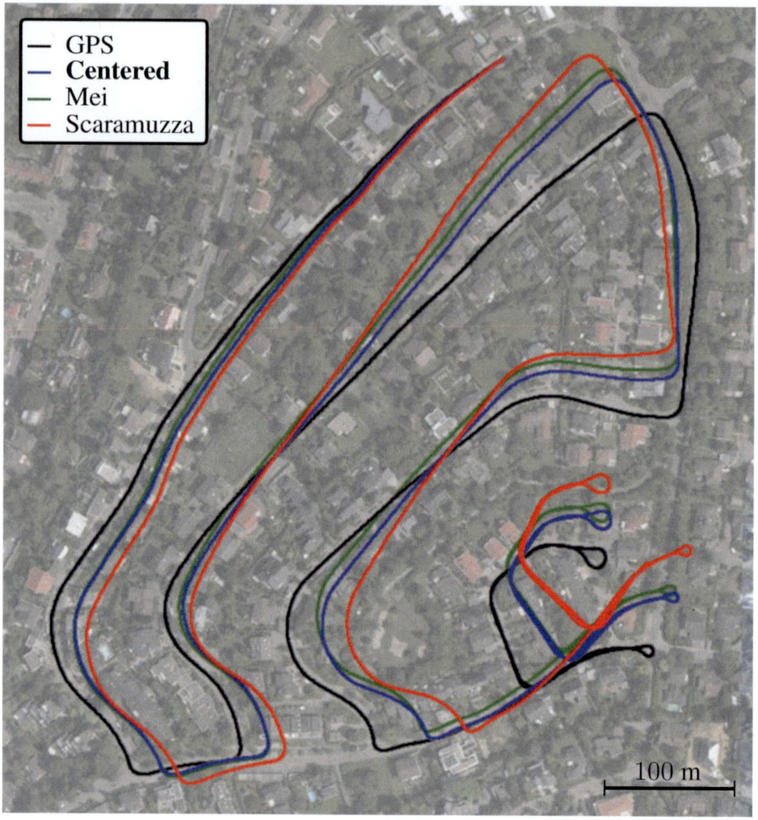

Figure 4.9.: **Ego-motion Estimation Both Cameras.** This figure shows the estimated trajectory of the third sequence (4 000 frames) by minimizing the reprojection error in both camera images with the centered projection model and the central reference models.

centered projection model than with state-of-the-art central projection models. Therefore, in Table 4.3 on the right side, we only show the end-point error $\mathbf{e}_i(j)$ for minimizing the motion parameters in two spatial camera images for $j = 1\,000$ frames. The estimated trajectories by minimizing in both images for two sequences are shown in Fig. 4.8 and Fig. 4.9 in top view overlaid on the BING satellite images.

In summary, the proposed centered projection model significantly improves the ego-motion estimation result with real catadioptric cameras which mainly deviate the single viewpoint condition in axial direction,

Error [m]	**Catadioptric**	Perspective
$\|\|\mathbf{e}_i\|\|$	2.84	8.28
e_{i_x}	1.70	2.67
e_{i_y}	1.84	3.37
e_{i_z}	0.50	5.96

Table 4.4.: **Comparison Catadioptric vs. Perspective.** This table shows the end-point errors $\mathbf{e}_i$ for ego-motion estimation with a stereo catadioptric camera setup compared to a stereo perspective camera setup by minimizing in both images. The errors are mean values over all frames while the end-point error is computed for $j = 200$ frames.

particularly in terms of altitude. Moreover, the computational cost of the centered model is less than for the *Scaramuzza* model and similar to the central *Mei* projection.

4.4.3. Motion Results compared to Perspective Stereo

We perform the same ego-motion estimation experiments without any motion constraint or bundle adjustment on perspective stereo images to show the benefit of using catadioptric cameras for an ego-motion application in difference to perspective cameras. We use the same approach as presented (Fig. 4.2) with a perspective projection model (see Appendix A.3) instead of the catadioptric projection models and minimize the reprojection error in both images. The bucketing and triangulation steps are adapted to perspective images with a bucketing in 16 squares and a direct triangulation from the rays through the image plane. To compare the motion estimation results of the different camera setups, we transform the perspective motion to the left catadioptric reference coordinate system, which means that for the perspective motion also the x-y-plane presents approximately the top view plane.

Naturally, the feature matches to compute the ego-motion are in a limited field of view and only in front of the ego-vehicle. This explains the larger end-point error as shown in Table 4.4, since the translation part is more difficult to estimate. The values in Table 4.4 are again the mean norm end-point error and the mean end-point errors in x-, y- and z-direction for $j = 200$ frames over all frames and sequences, as in the previous section for minimizing the errors in one camera image only.

Thus, we reduce the end-point error for motion estimation with catadioptric cameras and the proposed centered projection model approximately by factor three compared to perspective cameras. A resulting trajectory with

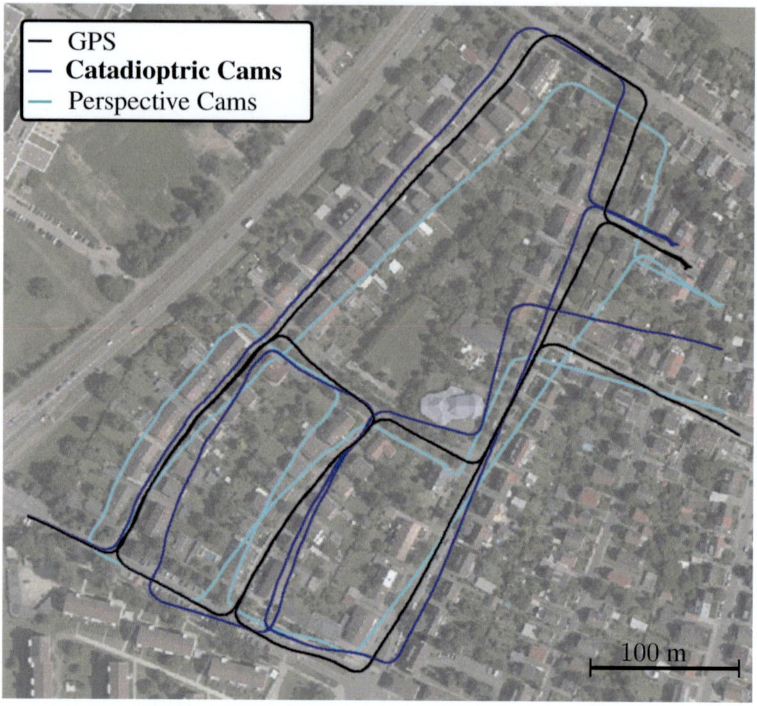

Figure 4.10.: **Ego-motion Catadioptric vs. Perspective.** This figure shows the estimated trajectories by minimizing the reprojection error in both camera images with a catadioptric stereo camera setup and a perspective stereo camera setup.

perspective cameras and catadioptric cameras in top view overlaid on the BING satellite image is shown in Fig. 4.10.

4.4.4. Top View Map

From the resulting accurate trajectories, we compute top view maps of the driven path and the nearby surrounding. To create a 2D top view map, we reproject the resulting 3D trajectory on the x-y-ground plane. A top view map is generated by stitching together generated birds-eye view images with the motion information. Therefore, we create a distortion free virtual perspective birds-eye view image from the centered catadioptric image.

In general, a point $\mathbf{p}_{vip}$ on a virtual perspective image plane is given by

$$x_{vip} = (f_p \sin\theta_0 - v_p \cos\theta_0)\cos\varphi_0 + u_p \sin\varphi_0 \quad (4.7)$$

$$y_{vip} = (f_p \sin\theta_0 - v_p \cos\theta_0)\sin\varphi_0 - u_p \cos\varphi_0 \quad (4.8)$$

$$z_{vip} = f_p \cos\theta_0 + v_p \sin\theta_0 \quad (4.9)$$

where u_p and v_p are the pixels on the image plane, φ_0 and θ_0 is the position of the origin from the virtual image, and f_p the virtual focal length in pixel. This relationship is denoted in Fig. 4.11a. For the top view image, we remap a virtual image parallel to the ground plane ($\varphi_0 = 0$, $\theta_0 = -180°$) with the virtual focal length $f_v = 200$ pixels. In Fig. 4.11 the remapped virtual image (c) and the corresponding original catadioptric image (b) are shown. By choosing another focal length, we achieve a different zoom factor of the perspective birds-eye view image.

Afterwards, we stitch the virtual images of every frame to obtain a high fidelity top view map of the driven path. Hence, we rotate each virtual image with respect to the first one and map it to the translated position. A resulting top view map with a driven path of 600 frames, computed with the frame-to-frame method and without any bundle adjustment or loop closure, is shown in Fig. 4.12. Fig. 4.13 presents two further parts of other top view maps, including lane markings on the street, computed from the visual estimated motion.

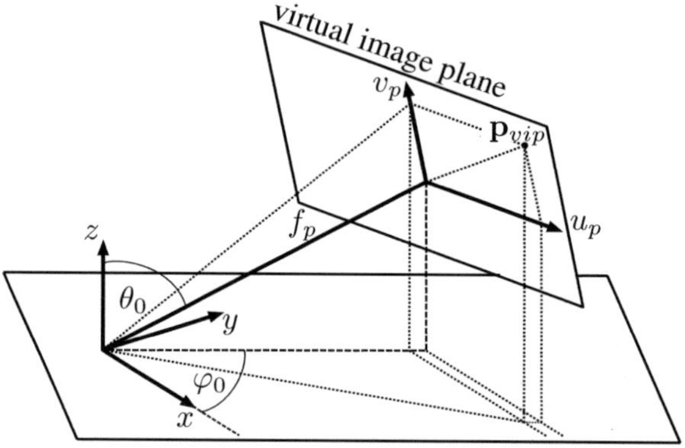

(a) Virtual Image Plane

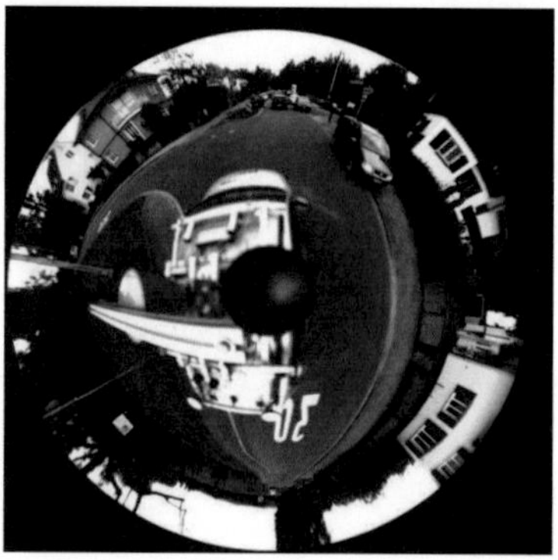

(b) Catadioptric Image

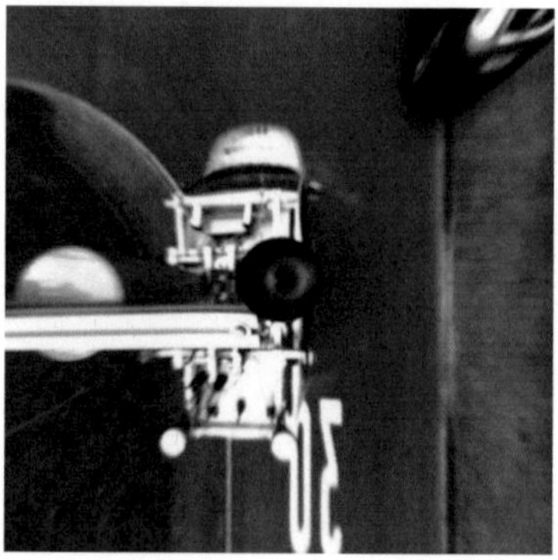

(c) Virtual Perspective Image

Figure 4.11.: **Virtual Perspective Birds-Eye View Image.** In (a) the relationship and the parameters to create a virtual perspective image from the centered catadioptric image are depicted. In (b) the original catadioptric image is shown from which we compute the virtual perspective birds-eye view image (c).

Figure 4.12.: **Top View Map.** This figure shows a driven trajectory of 600 frames estimated with the centered projection model in the computed top view map. The trajectory is entirely computed from two-frame motion with catadioptric camera images and depicted as green stars.

Figure 4.13.: **Top View Parts.** This figure shows two further parts of top view maps computed with the centered projection model including a more complex scenario with different lane markings.

Chapter 5

Dense 3D Reconstruction

This chapter presents a second application for autonomous vehicles which can be improved by the large field of view of catadioptric cameras and benefits from the proposed efficient and accurate projection model. We propose a novel method for dense 3D reconstruction of the static parts of the whole scene around the vehicle with a catadioptric stereo camera setup.

The chapter starts with an overview of existing 3D reconstruction methods for omnidirectional cameras. Afterwards, we explain the construction of one $360°$ disparity image from two consecutive frames of a stereoscopic catadioptric camera. We use planarity priors to improve the disparity image resulting in a smooth omnidirectional depth image. The results of our approach are compared against laser-based ground truth depth maps in different urban scenarios.

5.1. State-of-the-Art

Similar to visual ego-motion estimation, there is a large body of literature focusing on dense stereo matching or 3D reconstruction with perspective cameras. An overview and evaluation of different approaches for perspective cameras is given in [101, 41]. In general, existing methods can be divided in local methods, which compute a similarity measure in a small window, and global methods, which solve an optimization problem and incorporate smoothness priors.

However, there exists little work on dense 3D reconstruction with omnidirectional cameras. Some approaches use perfect vertically aligned catadioptric stereo systems which simplifies the epipolar lines to radial lines in the catadioptric images [48, 125]. Unfortunately, these systems typically only allow for accurate reconstruction in a very short range, due to a small baseline. Svoboda and Pajdla [113] describe the epipolar geometry for all kinds and positions of central catadioptric camera systems. They show that epipolar lines correspond to general conics in the omnidirectional image. For general camera motion, some authors propose to reproject the omnidirectional image to multiple perspective images, such as Gehrig [40], or to one panoramic image on a virtual cylinder, e.g., Bunschoten and Kröse [16] and Gonzalez and Lacroix [50]. Afterwards, stereo correspondences are established by searching along sinusoidal shaped epipolar curves [16, 64, 116] in the panoramic images.

Gonzalez and Lacroix [50] rectify the panoramic images resulting in epipolar curves reduced to straight lines. Thus, the correspondence search in panoramic images is simplified to a one-dimensional search problem. Geyer and Daniilidis [46] propose a conformal rectification method for parabolic images directly applied to the omnidirectional image. They remap the observations from bipolar coordinates to a rectangular grid to simplify the search problem. Another rectification method which we use in this work is the spherical rectification [34, 71, 5]. This rectification method is very flexible, can handle the existence of more than one epipole, and does not depend on a particular projection model.

Some approaches use two or multiple views to obtain a dense 3D reconstruction of the complete environment. Arican and Frossard [5] optimize a pixel-wise energy function using the graph-cut algorithm to compute a dense depth image from two rectified omnidirectional images. Similar approaches were presented by Fleck et al. [33] using three omnidirectional images and He et al. [54] using two panoramic images to obtain dense panoramic disparity images. Lhuiller [69] concentrates on the problem of fusing depth maps from multiple views for larger models from video sequences. Initially, they reconstructed the scene from three consecutive frames which are projected onto six faces of a virtual cube in order to allow for traditional stereo matching techniques. The local results are fused into a global model by selecting the most reliable viewpoint for each scene point and merging the 3D points using their median. This approach has been extended in [128] towards reconstruction of larger models from video sequences.

5.2. Dense 3D Reconstruction

In this work, we present a novel dense 3D reconstruction method for omnidirectional images which does not rely on constructing virtual perspective

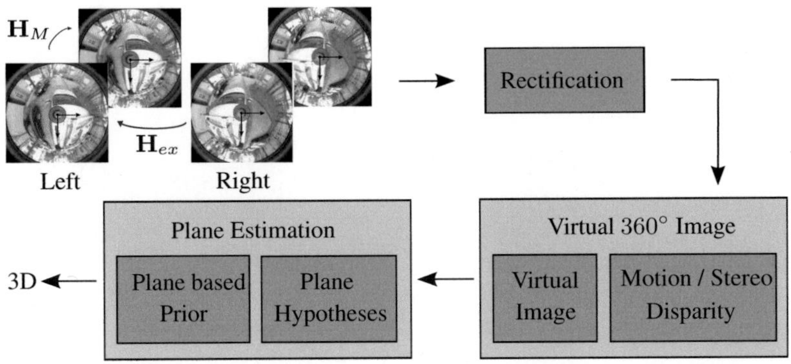

Figure 5.1.: **Dense 3D Reconstruction.** This figure shows an overview over the presented dense reconstruction method which is separated in three main parts: The rectification, the computation of a virtual panoramic image, and the plane estimation. The method based on the same camera setup presented in Fig. 4.1c and needs the result of the ego-motion estimation as well as the calibration and two consecutive stereo frames as input.

images from omnidirectional ones as in [40]. Moreover, we overcome the problem of depth blind spots induced from reconstructions with only two cameras as shown in Chapter 4.3.1. We eliminate this problem by using the same camera setup as for the ego-motion estimation (shown in Fig. 4.1c) with two catadioptric cameras at two consecutive frames. Thus, we obtain stereo information from four catadioptric image pairs (2×2 images), two spatial image pairs (spatial stereo) at time t and $t + 1$ as well as two temporal image pairs (motion / temporal stereo), one for the left and one for the right camera. Through combination of the catadioptric image pairs into one unified view, we enable efficient inference and overcome the problems of blind spots near the epipoles and occluded regions in some parts of the images. The basics of this approach have been published in [103].

The method to achieve a dense 3D reconstruction of the complete environment is divided into three main parts as shown in Fig. 5.1. The inputs to our method are two consecutive stereo image pairs, the intrinsic and extrinsic calibration (from Chapter 3.2) and the ego-motion estimation between two consecutive frames (from Chapter 4.3). In a first step, we rectify the four catadioptric image pairs to enable efficient scanline methods. To obtain one 360° virtual image, we compute a disparity image from each rectified image pair by applying efficient matching strategies and combine these four disparity images to one resulting panoramic depth map. Finally,

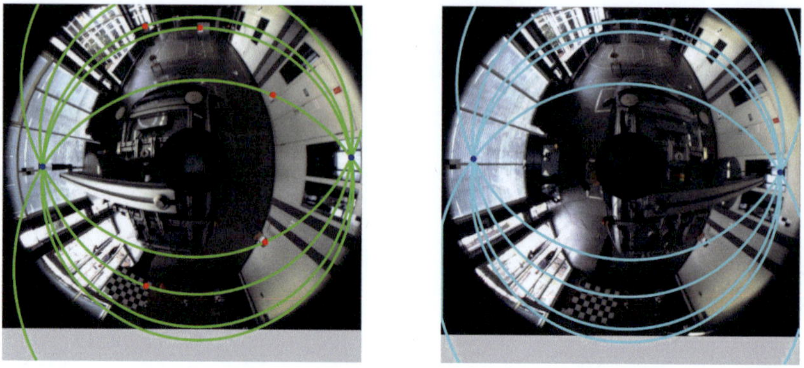

Figure 5.2.: **Epipolar Geometry.** This figure shows the epipolar conics in a catadioptric image pair. The cyan conics in the right image depict the corresponding curves to the red dots in the left image. The blue dots, where all curves intersect, denote the epipoles.

we consider the static parts of the scene to follow the augmented manhattan world assumption [102], which means that the scene can be described by vertical and horizontal planes in 3D. We estimate plane hypotheses with a novel voting scheme for 3D planes in omnidirectional space and obtain an omnidirectional depth map by selecting the best plane hypothesis for each pixel solving a discrete energy minimization problem.

5.2.1. Rectification

A rectification process is necessary for an efficient dense 3D reconstruction to reduce the computational cost and simplify the correspondence search to a one-dimensional problem. Therefore, the images are remapped such that corresponding points are located on the same pixel row in a rectified image pair. For perspective image pairs corresponding points lie on straight lines in the images. All these lines in one image intersect in one point, the epipole. For common perspective stereo rigs the epipoles are located outside the image plane. In difference corresponding points in central catadioptric image pairs lie on conics [113]. Moreover, there exist two epipoles in each image plane where all conics intersect each other as shown in Fig. 5.2. The cyan colored conics in the right image correspond to the red dots and green conics in the left image. The blue dots denote the two epipoles in each image. Since there exist two epipoles in the image plane, standard rectification methods used for perspective cameras which are based on homographies [53] cannot be applied. Besides, the description of the epipolar geometry as well as an efficient remapping of the images is only valid for central catadioptric cameras which fulfill the single viewpoint condition.

As we apply the centered model which simplifies each non-central model by a central one, we are able to compute the epipolar geometry and remap the omnidirectional images to a rectified image pair. We use the spherical rectification similar to [34, 5] in the spherical domain with the 3D world point $\mathbf{p} = [\rho, \varphi, \theta]^\mathsf{T}$ with

$$\varphi = \arctan \frac{y}{x} \qquad \theta = \arctan \frac{\sqrt{x^2 + y^2}}{z} \qquad \rho = \sqrt{x^2 + y^2 + z^2} \tag{5.1}$$

computed from the world point $\mathbf{p} = [x, y, z]^\mathsf{T}$ in the Cartesian coordinate system. The relationship between Cartesian and spherical coordinates is depicted in Fig. 2.9. The epipolar constraint in the spherical domain reads as

$$\mathbf{m}_1^\mathsf{T} \mathbf{E}_{12} \mathbf{m}_2 = 0 \quad \text{with} \quad \mathbf{E}_{12} = [\mathbf{t}]_\times \mathbf{R} \tag{5.2}$$

and is the same as for perspective cameras, with the difference that $\mathbf{m}_1$ and $\mathbf{m}_2$ are the three-dimensional projections in the first and second camera of the world point $\mathbf{p}$ on the mirror surface and not the projected two-dimensional image points. $\mathbf{E}_{12}$ is the essential matrix depending on the transformation $\mathbf{R}$ and $\mathbf{t}$ between the two frames of the image pair. The four epipoles $\mathbf{e}_{11}, \mathbf{e}_{12}, \mathbf{e}_{21}, \mathbf{e}_{22}$ on the unit sphere can be obtained from the essential matrix using the singular value decomposition [53].

After computing the symmetric epipoles on the unit sphere, the spherical images are rotated such that the epipoles coincide with the coordinate poles (z-axis). Thus, the line connecting both camera centers is the same as the new z-axis of the rotated coordinate system shown in Fig. 5.3. The remaining degree of freedom of this rotation is chosen to keep the remaining axis of the rotated coordinate system similar to the one of the original mirror coordinate system, which depends on the transformation between the two cameras. Thereafter, in the rotated spherical coordinate system a 3D world point $\mathbf{p}_S$ lies on the epipolar plane Π_R with the same azimuth angle φ_S in both rotated coordinate systems. Thus, epipolar great circles coincide with the longitudes and disparity estimation reduces to a one-dimesional search problem with constant azimuth angle φ_S. The rectified spherical image depends on the azimuth angle $\varphi_S \in [0, 2\pi]$ and the inclination angle $\theta_{S_i} \in [0, \pi]$ from the rotated world points

$$\mathbf{p}_S = (x_S, y_S, z_S)^\mathsf{T} = \mathbf{R}_{S_i}^{-1} \cdot \mathbf{p}, \tag{5.3}$$

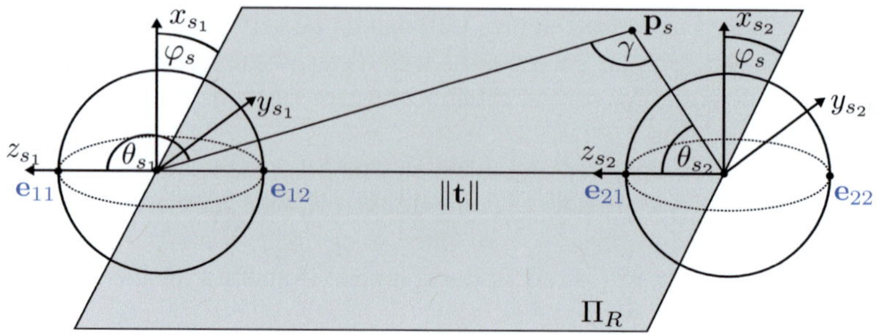

Figure 5.3.: **Spherical Rectification.** After applying the rectifying rotation, a 3D point $\mathbf{p}_S$ lies on the epipolar plane Π_R with the same azimuth angle φ_S in both rotated spherical coordinate systems. The rotated coordinate system depends on the relative position of both cameras determined by extrinsic calibration (spatial stereo) or motion estimation (motion stereo), respectively.

where $\mathbf{R}_{S_i}$ is the rotation matrix between the original and the rotated coordinate system. The index $i \in \{1, 2\}$ for θ_S and $\mathbf{R}_S$ denotes the first and the second camera of the image pair.

In this work, we have spatial as well as temporal image pairs. The relative positions between the cameras are determined from the extrinsic calibration $\mathbf{H}_{ex}(\mathbf{R}_{ex}, \mathbf{t}_{ex})$ and the ego-motion estimation $\mathbf{H}_M(\mathbf{R}_M, \mathbf{t}_M)$, respectively. The extrinsic calibration does not change during a sequence of images which means that the rectification map in the spatial case is computed once at the beginning during the calibration process. Since the transformation for the temporal case changes for every frame, this rectification map has to be computed at runtime. The rotation matrices $\mathbf{R}_{S_1}$ and $\mathbf{R}_{S_2}$ for the first and the second image of an image pair in the spatial stereo case are obtained as

$$\begin{aligned} \mathbf{R}_{S_1} &= [\mathbf{r}_1, \mathbf{r}_2, \mathbf{e}_{11}] \\ \mathbf{r}_2 &= \mathbf{y}_0 - (\mathbf{e}_{11}^T \mathbf{y}_0)\mathbf{e}_{11} \qquad\qquad \mathbf{R}_{S_2} = \mathbf{R}_{ex}\mathbf{R}_{S_1} \\ \mathbf{r}_1 &= \mathbf{r}_2 \times \mathbf{r}_3 \end{aligned}$$

where $\mathbf{y}_o$ denotes the normalized unit vector in y-direction of the original omnidirectional reference camera coordinate system (before rotation) and $\mathbf{e}_{11}$ is the epipolar point as illustrated in Fig. 5.3. For the temporal stereo case the direction of the translation is approximately parallel to the y-axis. Therefore, to avoid numerical instabilities for the image pair in temporal direction, we use $\mathbf{r}_2 = \mathbf{x}_0 - (\mathbf{e}_{11}^T \mathbf{x}_0)\mathbf{e}_{11}$ where $\mathbf{x}_o$ denotes the normalized

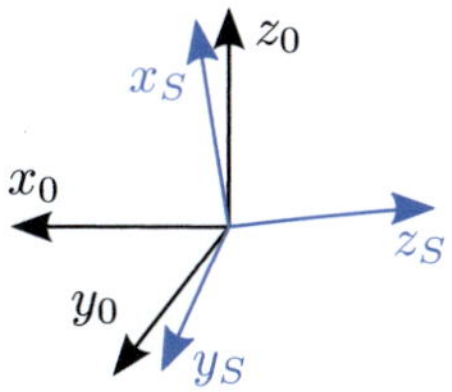

(a) Spatial Stereo

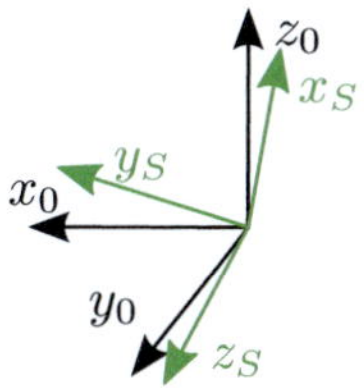

(b) Temporal Stereo

Figure 5.4.: **Rotated Coordinate Systems.** This figure shows the reference mirror coordinate systems (black) and the rotated sphere coordinate systems (colored) for the spatial stereo (a) and the temporal stereo case (b).

unit vector in original x-direction. The coordinate systems before and after the rotation are shown in Fig. 5.4 for the spatial stereo case (a) and for the temporal stereo case (b).

The result of the spherical rectification for a spatial image pair at one time step (a) and the result for a temporal image pair from the left camera (b) are depicted in Fig. 5.5. The horizontal lines show scanlines with the same azimuth angle φ_S on which corresponding points in both images are located. In case of temporal stereo, we only rectify a part of the image. For the left temporal case we rectify the region inside the magenta colored box, since the parts on the other side of the vehicle are better visible in the right temporal image pair. Consequently, we rectify a similar part for the right image pair. This limitation speeds-up the computation of the rectification maps for the temporal image pairs which needs to be done at runtime. We use bilinear interpolation to achieve smooth rectified image pairs. Note, this rectification method is also valid for other central image models. In Fig. 5.6 we show the result of the spherical rectification for a temporal fisheye stereo image pair with the same scanlines as in the catadioptric case.

5.2.2. Virtual Panoramic Image

After spherical rectification, we compute one disparity image for each rectified image pair using Semi-Global Matching [56] which has shown good performance for perspective image pairs [41]. In Fig. 5.7 the first images of the four rectified input image pairs are shown for each case. In (a) and (b) the spatial stereo images for two consecutive frames are depicted, and (c) and (d) show the rectified parts of the temporal image pairs of the left and right camera. Moreover, the figure shows the resulting disparity images (e) - (h) for the corresponding image pairs.

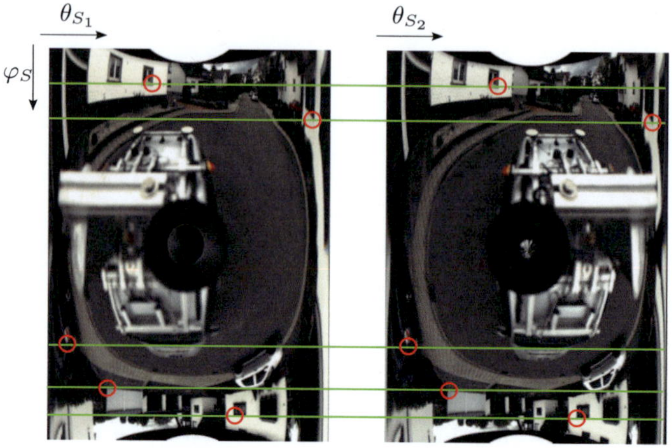

(a) Rectified Spatial Stereo Image Pair

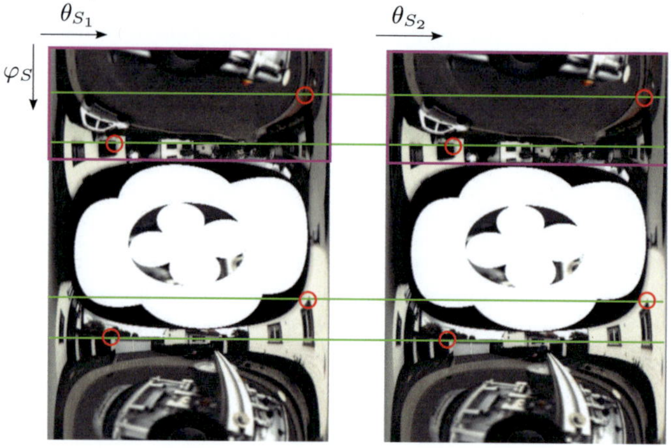

(b) Rectified Temporal Stereo Image Pair

Figure 5.5.: **Rectified Catadioptric Stereo Pairs.** This figure shows the result of the spherical rectification for a spatial (a) and a temporal (b) stereo image pair. The horizontal green lines are scanlines with the same azimuth angle φ_S on which corresponding points are located. The red circles show corresponding points in both images. For the temporal image pair we later use only a part of the rectified image denoted with the magenta box for efficiency.

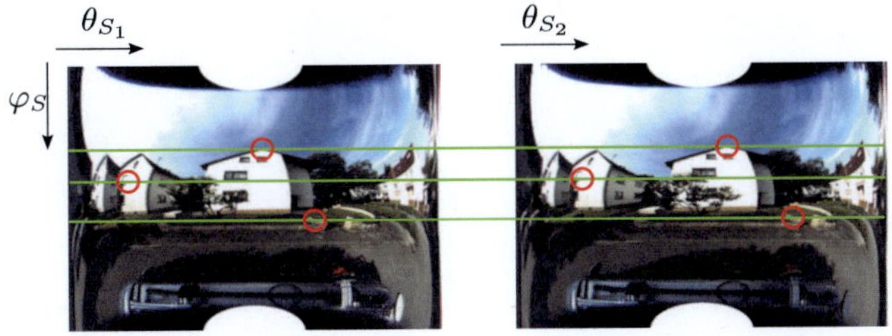

Figure 5.6.: **Rectified Fisheye Image Pair.** This figure shows the result of the rectification process for a temporal fisheye stereo image pair. The red dots denote again corresponding points and the green lines depict horizontal scanlines.

As already shown in Chapter 4.3.1 the accuracy of the 3D reconstruction depends on the position of the resulting 3D point. Reconstructing points along the baseline of the cameras is not possible. In the rectified images this fact leads to highly distorted images near the epipoles which are located at the margins ($\theta_{S_i} = 0$ and $\theta_{S_i} = \pi$) of the rectified images. Therefore, we clip the margins of the images. For the spatial stereo disparity images this means that we extract 3D information only in the front (120°) and rear (120°) parts of the ego-vehicle. For the motion stereo disparity we extract 3D information only from the corresponding side (120° each) of the camera. Furthermore, large parts of the spherically rectified images are occluded by the recording platform itself. Therefore, we overlay a mask of the ego-vehicle where we do not compute disparity.

In the spherical domain, the angular disparity is defined as

$$\gamma = |\theta_{S_1} - \theta_{S_2}|, \tag{5.4}$$

the difference between the angles θ_{S_1} and θ_{S_2} of the two viewing rays from the two images corresponding to the 3D world point $\mathbf{p}_S$. For calculating the depth ρ_S of $\mathbf{p}_S$ two cases have to be distinguished as illustrated in Fig. 5.8. We differentiate the case where the camera moves in negative z-direction (a) and the case where the camera moves in positive z-direction (b). In the first case we have

$$\sin\gamma = \frac{h}{\rho_S} \quad \text{and} \quad \sin\theta_{S_2} = \frac{h}{\|\mathbf{t}\|}. \tag{5.5}$$

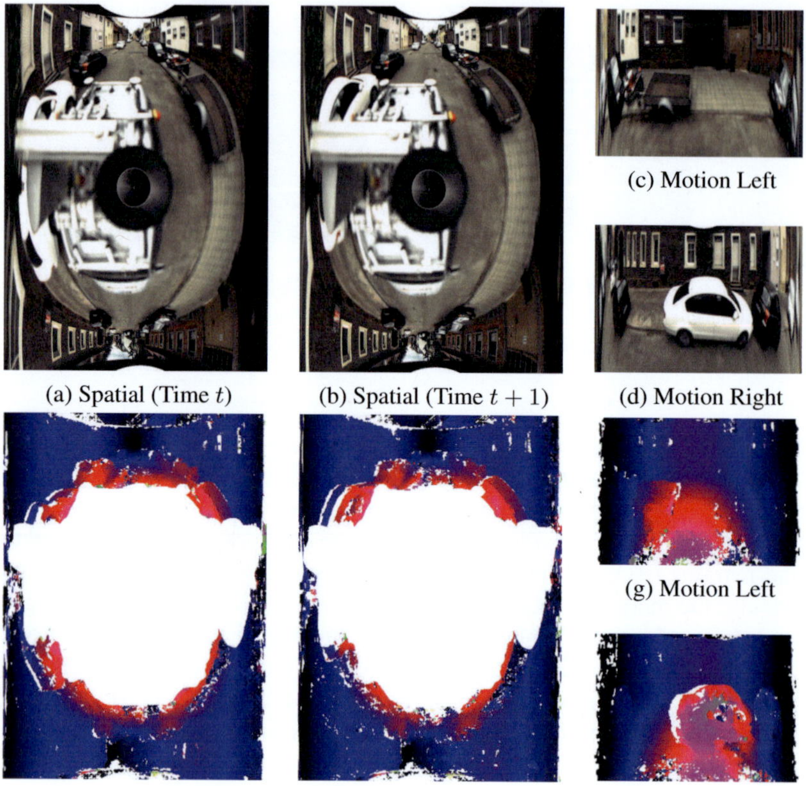

(a) Spatial (Time t) (b) Spatial (Time $t+1$) (c) Motion Left (d) Motion Right

(e) Spatial (Time t) (f) Spatial (Time $t+1$) (g) Motion Left (h) Motion Right

Figure 5.7.: **Rectified and Disparity Images.** This figure shows the rectified images (a)-(d) as well as the corresponding resulting disparity images (e) - (h) obtained by Semi-Global Matching for all four image pairs.

Thus, the depth ρ_S is computed to

$$\rho_S = \frac{\|\mathbf{t}\| \cdot \sin \theta_{S_2}}{\sin \gamma} \tag{5.6}$$

where $\|\mathbf{t}\|$ denotes the length of the baseline between the cameras. Similarly, for the second case with camera motion in positive z-direction, we obtain

$$\rho_S = \frac{\|\mathbf{t}\| \cdot \cos \theta_{S_2}}{\sin \gamma}. \tag{5.7}$$

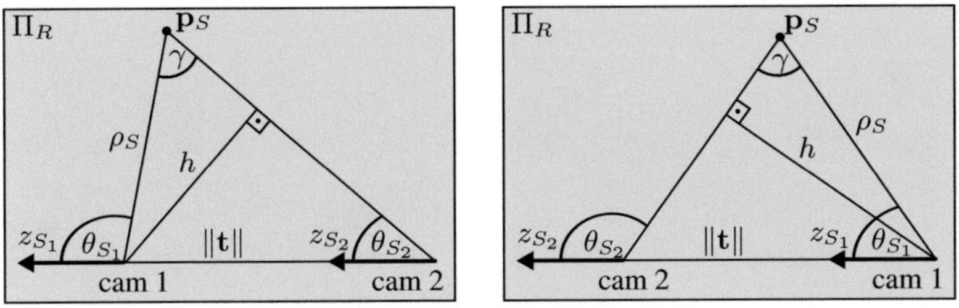

(a) Camera moves in negative z-direction (b) Camera moves in positive z-direction

Figure 5.8.: **Reconstruction.** This figure shows the derivation of the depth ρ_S as a function of γ and θ_{S_2} for the two distinguished cases with camera motion in negative z-direction (a) and in positive z-direction (b).

With the depth ρ_S we compute the 3D point $\mathbf{p}$ in the original Cartesian coordinate system

$$\mathbf{p} = \mathbf{R}_{S_1} \begin{pmatrix} \rho_S \sin\theta_S \cos\varphi_S \\ \rho_S \sin\theta_S \sin\varphi_S \\ \rho_S \cos\theta_S \end{pmatrix} . \tag{5.8}$$

After estimating the 3D points from each of the four spherical disparity images, we combine all information in a single new virtual 360° intensity and depth image. The coordinate system of the new virtual camera is chosen to be in the center between the four original camera centers, namely between the left and right camera at two consecutive frames as shown in Fig. 5.9. This coordinate system is chosen to minimize the relative displacements of all reflected rays from all camera pairs. We transform all 3D points to the new coordinate system by first transforming them to the previous left coordinate system, which is chosen as the reference camera coordinate system. Next, we shift the 3D points by a pure translation $\mathbf{t}_V$ to the new coordinate system in the middle between all four cameras. Furthermore, the x-y-plane of the new virtual camera coordinate system is chosen parallel to the ground plane. We estimate this rotation $\mathbf{R}_V$ by computing the dominant plane below the camera using RANSAC plane fitting of all 3D points.

While merging the four disparity images we are faced with overlapping regions in 3D space despite already cropped disparity images. Depth values in the overlapping regions are merged by computing the mean value. This

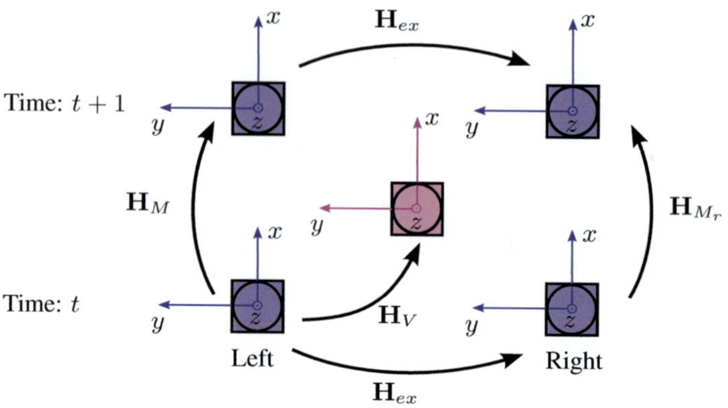

Figure 5.9.: **Virtual Camera Coordinate System.** This figure shows the position of the virtual camera coordinate system between the left and right camera at two consecutive frames.

is sufficient since the reconstruction accuracy in the remaining disparity images is similar over all parts. Through remapping all 3D points to the new virtual camera coordinate system with

$$\mathbf{H}_V = \begin{bmatrix} \mathbf{R}_V & \mathbf{t}_V \\ \mathbf{0} & 1 \end{bmatrix}, \tag{5.9}$$

we obtain one single virtual 360° intensity image $I(\varphi_V, \theta_V)$ and an inverse depth image $D(\varphi_V, \theta_V)$. The inverse depth is defined as

$$D = 1/r \quad \text{with} \quad r = \sqrt{x_V^2 + y_V^2}$$

independent of the z-component of each 3D point. Both virtual 360° images, the intensity and inverse depth image, where the color denotes the depth, are illustrated in Fig. 5.10.

5.2.3. Plane Estimation

After computing the virtual 360° inverse depth image, we improve the quality of the depth image to achieve a smoother 3D reconstruction and reject remaining outliers. We propose this step since catadioptric images, in particular, suffer from blur and low contrast. To improve the depth images, we describe the static part of the 3D world with horizontal and vertical planes

Figure 5.10.: **Virtual Panoramic Images.** These images show the virtual 360° intensity image (top) as well as the virtual 360° inverse depth image (bottom), where the color denotes the inverse depth, computed from the four image pairs.

following the augmented manhattan world assumption [102]. This assumption does not require vertical planes to be orthogonal with respect to each other as in [35, 36] but only with respect to the horizontal planes. Indoor scenarios are often composed of mainly vertical and horizontal planes as in [107, 129], but also many urban scenes follow this assumption. In difference to perspective cameras, planes in 3D do not correspond to planes in the catadioptric image. Therefore, we cannot use planarity priors which have been proposed for stereo matching with traditional perspective cameras [76, 124].

We present a simple representation for vertical and horizontal planes in catadioptric images. To find plane hypotheses in the image, we first partition the virtual 360° image into approximately 1 000 superpixels using the recently proposed StereoSLIC [124] algorithm. Thus, we reduce the number of pixels for which plane hypotheses are estimated for efficient inference. This partitioning is illustrated in Fig. 5.11 for one virtual intensity image. After computing plane hypotheses for the whole image, we estimate the best plane hypothesis for each superpixel. We formulate the problem as a discrete energy minimization problem and solve it using belief propagation.

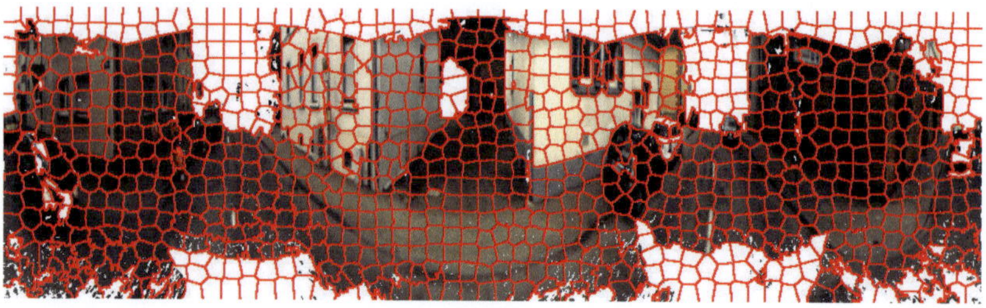

Figure 5.11.: **Superpixels.** This figure depicts the superpixel partitioning computed with the StereoSLIC algorithm on the virtual 360° intensity image.

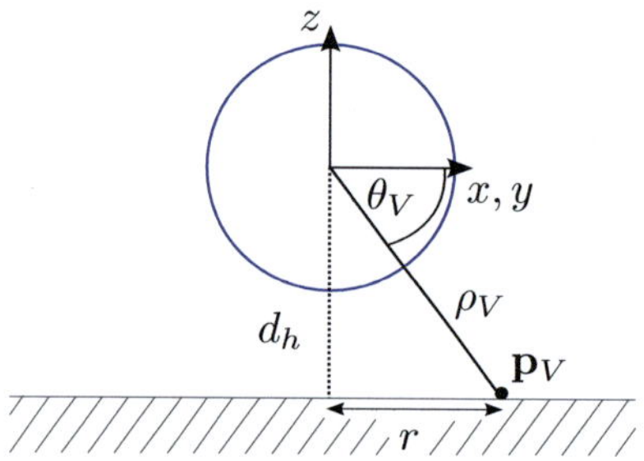

(a) Horizontal Plane (Side View)

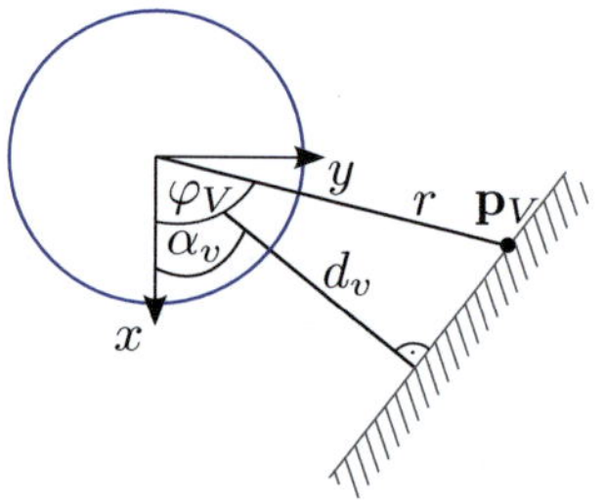

(b) Vertical Plane (Top View)

Figure 5.12.: **Plane Hypotheses.** This figure shows the relationship between a point $\mathbf{p}_V$ described by the spherical parameters φ_V, θ_V, and its depth r and the plane parameters d_h, d_v, and α_v for horizontal (a) and vertical planes (b) in the coordinate system of the virtual camera.

Plane Hypotheses For the description of vertical and horizontal planes we use the fact that the coordinate system of the virtual 360° image is parallel to the ground plane (x-y-plane). Thus, we are able to describe horizontal planes which are parallel to the ground plane with a single variable (distance d_h) as depicted in Fig. 5.12a. Vertical planes which are perpendicular to the ground plane are described with two variables (angle α_v and distance d_v) as illustrated in Fig. 5.12b. Since the depth r from the inverse depth image $D(\varphi_V, \theta_V) = 1/r$ is independent from the z-component of a 3D point, the relationship between a 3D point $\mathbf{p}_V$ and the distance of a horizontal d_h and vertical plane d_v passing through the point $\mathbf{p}_V$ are given by

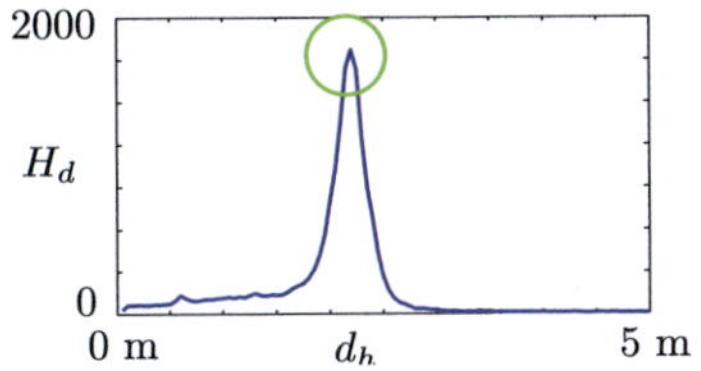

(a) Horizontal Hough Array $H_h(d_h)$

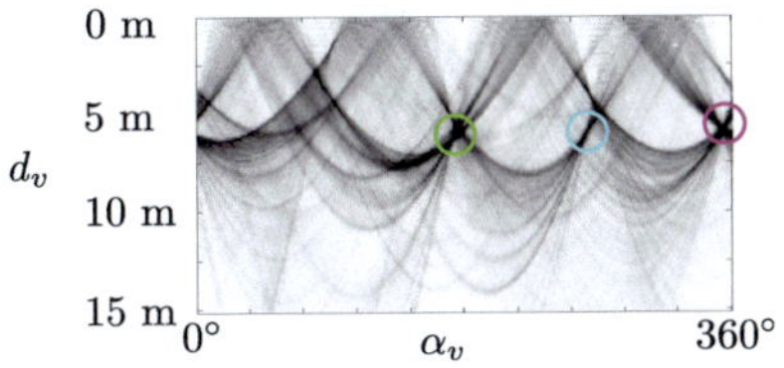

(b) Vertical Hough Array $H_v(d_v, \alpha_v)$

(c) Images with a Horizontal Plane

(d) Images with three Vertical Planes

Figure 5.13.: **Hough Voting for Planes in Omnidirectional Depth Images.** The images show the results of the hough transformation for horizontal (a) and vertical planes (b). In (c) and (d) the virtual panoramic intensity (top) and inverse depth image (bottom) are shown with the planes corresponding to the maxima in the hough space overlaid. The different colors in the vertical case show different plane hypotheses corresponding to the same colored maxima.

$$d_h(r, \theta_V) = r \cdot \tan \theta_V \tag{5.10}$$

$$d_v(r, \varphi_V, \alpha_v) = r \cdot \cos(\varphi_V - \alpha_v) \tag{5.11}$$

where φ_V and θ_V denote the pixel position in the inverse depth image $D(\varphi_V, \theta_V)$.

This description of the planes suggests a simple hough voting scheme to estimate all vertical and horizontal planes which exist in the virtual image, similar to the hough transformation for extracting lines. We use a one-dimensional horizontal plane accumulator array $H_h(d_h)$ as shown in Fig. 5.13a to accumulate the votes for horizontal planes over all superpixels in the virtual panoramic inverse depth image. For vertical planes we use a two-dimensional vertical plane accumulator array $H_v(d_v, \alpha)$ as depicted in Fig. 5.13b. We disambiguate pixels belonging to horizontal and vertical surfaces to make the votes more discriminative by casting each vote with an additional weight. The likelihood of a pixel belonging to a horizontal or vertical plane is used as weight function. This likelihood is modeled by logistic regression using the vertical inverse depth image gradient as input. The parameters of the sigmoid function are estimated using a representative training image for which all horizontal and vertical planes have been labeled manually.

We compute the maxima of the horizontal $H_h(d_h)$ and vertical $H_v(d_v, \alpha)$ accumulator arrays using an efficient non-maxima suppression implementation [84] slightly extended to handle panoramic cycle images. Fig. 5.13 shows the resulting maxima in the horizontal (a) and vertical (b) accumulator array as well as the corresponding pixels belonging to the horizontal (c) and vertical plane hypotheses (d) are depicted as colored pixels in the virtual inverse depth and virtual intensity images. For the vertical case we show three maxima corresponding to three different plane hypotheses in the hough space in different color. The superpixels which belong to the corresponding planes are identically colored in the virtual images. On average, we get 2.3 horizontal plane hypotheses and 46 vertical plane hypotheses per image depending on the threshold for non-maxima suppression.

Plane Optimization Given the plane hypotheses we find the best plane hypothesis for each superpixel under the assumption that nearby superpixels likely belonging to the same surface. We formulate the problem of assigning each superpixel to one of the plane hypotheses as a discrete energy minimization problem. The minimized energy function is

$$E(\mathcal{S}) = \sum_{s \in \mathcal{S}} \underbrace{[E_{u_1}(s) + E_{u_2}(s)]}_{\text{unary terms}} + \sum_{(s_1,s_2) \in \mathcal{N}_S} \underbrace{E_p(s_1, s_2)}_{\text{pairwise terms}} \tag{5.12}$$

with unary terms E_u and pairwise terms E_p. The variables of interest, each corresponding to one superpixel is denoted as $\mathcal{S} = \{s_1, \ldots, s_M\}$, where s takes a discrete plane index $s \in \{1, \ldots, N\}$ as value. Here, M denotes the total number of superpixels in the image and N is the number of plane hypotheses, while $\mathcal{N}_S$ denotes the set of neighboring superpixels, i.e., all superpixels that share a common boundary.

The first unary energy term models the inverse depth fidelity

$$E_{u_1}(s) = w_{u_1} a(s) \sum_{\mathbf{q}_V \in \mathcal{Q}_s} \left[\epsilon_u(\underbrace{\hat{D}(\mathbf{q}_V, s) - D(\mathbf{q}_V)}_{x_{u_1}}) \right] \tag{5.13}$$

with weight parameter w_{u_1}. Here, $\hat{D}(\mathbf{q}_V, s)$ is the inverse depth at pixel $\mathbf{q}_V = (\varphi_V, \theta_V)^\mathsf{T}$ predicted from the plane with index s and $D(\mathbf{q}_V)$ is the inverse depth estimate at pixel $\mathbf{q}_V$ from the virtual inverse depth image. The function $\epsilon_u(x_{u_1}) = \min(|x_{u_1}|, \tau_u)$ is a robust penalty function with truncation parameter τ_u. Furthermore, $\mathcal{Q}_s$ denotes the set of all pixels with valid inverse depth hypothesis $D(\mathbf{q}_V)$ which are covered by superpixel s. The function $a(s) \in [0, 1]$ predicts the accuracy of the inverse depth map D averaged over superpixel s from training data. We introduce this function as we found the reliability of Semi-Global Matching to correlate strongly with image blur and hence also image location when dealing with omnidirectional images. In practice, we take $a(s)$ as the average ratio of correctly predicted depth values computed from a held-out training set.

The second unary term models the prior probability for surfaces to be horizontal or vertical and is given by

$$E_{u_2}(s) = w_{u_2} \times \begin{cases} 2\,p_h(s) - 1 & \text{if } s \in \mathcal{H} \\ 1 - 2\,p_h(s) & \text{otherwise} \end{cases} \tag{5.14}$$

where w_{u_2} is a weight parameter and $\mathcal{H}$ is the set of horizontal planes with

$$p_h(s) = \frac{1}{|\mathcal{Q}_s|} \sum_{\mathbf{q}_V \in \mathcal{Q}_s} p'_h(\mathbf{q}_V) \in [0, 1] \tag{5.15}$$

is the prior probability of superpixel s being horizontal, where $p'_h(\mathbf{q}_V)$ is the probability of pixel $\mathbf{q}_V$ being horizontal. We compute this probability from a separate training set augmented with manually labeled polygons of vertical and horizontal surfaces. For plane hypotheses that agree with the expected plane type, Eq. 5.14 assigns a positive score and otherwise a negative score.

Our pairwise model encourages neighboring superpixels to agree at their boundaries

$$E_p(s_1, s_2) = w_p \sum_{\mathbf{q}_V \in \mathcal{B}_{s_1,s_2}} \epsilon_p(\underbrace{\hat{D}(\mathbf{q}_V, s_1) - \hat{D}(\mathbf{q}_V, s_2)}_{x_p}) \tag{5.16}$$

where w_p is a smoothness parameter and $\mathcal{B}_{s_1,s_2}$ is the set of boundary pixels that are shared between superpixel s_1 and s_2. Similar to the depth fidelity term, we take $\epsilon_p = \min(|x_p|, \tau_p)$ as the robust penalty function with truncation parameter τ_p.

We use min-sum loopy belief propagation [12] to approximately minimize the energy function and select the best plane for each superpixel. The parameters of the energy model are estimated using a separate training sequence with 80 images with labeled ground truth information. We use Bayesian optimization [62] to estimate the parameters from the training data since Eq. 5.12 depends non-linearly on the parameters τ_u and τ_p, yielding in our case $w_{u_1} = 1.2$, $w_{u_2} = 1.0$, $w_p = 1.0$, $\tau_u = 0.05$, and $\tau_p = 0.08$.

5.3. Evaluation

We evaluate the dense 3D reconstruction approach on stereo sequences captured with the sensor setup proposed in Section 4.2. We show quantitative and qualitative results for different urban scenarios. For the quantitative results we compare against laser-based ground truth.

5.3.1. Ground Truth

For the quantitative comparison we use the Velodyne laser scanner as a reference sensor. We captured a dataset with 152 diverse and challenging urban scenes. The dataset is divided in 80 training and 72 test scenes. To evaluate the quantitative results we focus on static scenes without any moving parts. This allows us to accumulate the laser point cloud (± 5 frames) with an ICP point-to-plane fitting to achieve dense ground truth maps. By remapping the 3D laser points to the virtual camera coordinate system and to a panoramic image, we obtain a virtual inverse depth image with laser-based ground truth depth. The panoramic image with the depth from the Velodyne is shown in Fig. 5.16a (top).

Note that the vertical field of view of the Velodyne is smaller than the field of view of the catadioptric camera. For the quantitative evaluation we only consider image parts where image and laser information is provided. We have also manually labeled all horizontal and vertical planes in the images to evaluate the quality of depth information depending on surface inclination. The presented Hough transformation and the ground truth depth information from the laser scanner is used to determine the parameters of the labeled planes.

5.3.2. Quantitative Results

We evaluate the presented dense 3D reconstruction against state-of-the-art stereo vision algorithms which have shown good performance on standard perspective stereo tasks [41]. We apply simple Block Matching (BM), Semi-Global Matching (SGM) [56], as well as the recently proposed StereoSLIC algorithm [124] on the omnidirectional images. For Block Matching and Semi-Global Matching we use the OpenCV implementation. We also implement a simple winner takes all (WTA) plane selection strategy for the proposed approach as a reference method, which selects the best plane independently for each superpixel. This plane selection strategy investigates the importance of the proposed plane-based prior. The algorithm corresponds to minimizing Eq. 5.12 while ignoring the pairwise energies $E_p(s_1, s_2)$ and the horizontal prior $E_{u_2}(s)$.

To compare the results against each other, we compute an inverse depth error

$$e = |D_{gt}(\mathbf{q}) - D_{est}(\mathbf{q})| \tag{5.17}$$

for each pixel $\mathbf{q}$ for which ground truth is available, since the inverse depth error is independent from the distance to the measured points. Here, $D_{gt}(\mathbf{q})$ is the inverse depth in the Velodyne ground truth image at pixel $\mathbf{q}$ and $D_{est}(\mathbf{q})$ is the estimated inverse depth using the respective method. We fill in missing values in the estimated resulting inverse depth image using background interpolation [41, 56] to guarantee a fair comparison.

We report the mean number of bad pixels and the mean inverse depth error averaged over the full test set. As bad pixels we consider all pixels with an inverse depth error e larger than $0.05\ \text{m}^{-1}$. In Table 5.1 the mean percentage of bad pixels for all baseline methods and the proposed method averaged over the 72 test images is shown. Table 5.2 depicts the mean values of the inverse depth error for all methods. In both tables the first column states the errors for all pixels where depth ground truth is available. The other columns consider planar regions (vertical and/or horizontal) only. For the winner takes all algorithm we vary the threshold of the non-maxima suppression stage between 50 and 500 (denoted as WTA 50 / WTA 500 in the tables). Thereby, we achieve between 5 and 150 plane hypotheses for the winner takes all algorithm. For the proposed planarity prior method we set the non-maxima suppression threshold to 150.

The experiments show that the proposed plane-based method is able to achieve high-quality omnidirectional depth maps and outperforms state-of-the-art depth estimation techniques in terms of the 3D reconstruction error. The difference is especially pronounced for horizontal planes where we reduce the number of bad pixels as well as the mean inverse depth error. Besides, the proposed method also decreases the number of bad pixels for vertical planes. In Fig. 5.14 the depth error with the presented plane-based method is shown which depends exponentially from the measured distance. The green line shows the mean inverse depth error $e = 0.013\ \text{m}^{-1}$ computed depending on the distance and the red boxes depict the mean measured errors for the different distance ranges. For instance, the mean recon-

Bad Pixels (%)	All Pixel	All Planes	Horizontal Planes	Vertical Planes
SGM	11.89	13.41	17.45	2.52
BM	9.52	7.27	6.75	5.81
StereoSLIC	8.95	9.50	12.24	1.85
WTA 50	11.62	13.22	17.48	2.11
WTA 100	11.63	13.16	17.40	2.10
WTA 150	11.59	12.85	17.04	2.20
WTA 200	11.62	12.28	16.33	2.33
WTA 300	12.63	11.96	15.29	6.64
WTA 500	14.66	11.98	13.28	14.10
Ours	**4.04**	**1.24**	**1.03**	**1.51**

Table 5.1.: **Bad Pixels.** This table shows the mean percentage of bad pixels ($e > 0.05\ \mathrm{m}^{-1}$) for all baseline methods and the proposed method averaged over all 72 test images. The first column depicts the errors for all pixels where depth ground truth is available, while the other columns consider planar regions (of a specific type) only.

Mean Error ($\mathbf{m}^{-1}$)	All Pixel	All Planes	Horizontal Planes	Vertical Planes
SGM	0.026	0.029	0.034	**0.008**
BM	0.022	0.022	0.023	0.013
StereoSLIC	0.021	0.022	0.026	**0.008**
WTA 50	0.029	0.033	0.038	**0.008**
WTA 100	0.029	0.033	0.038	**0.008**
WTA 150	0.029	0.032	0.037	0.009
WTA 200	0.029	0.031	0.036	0.009
WTA 300	0.030	0.030	0.034	0.016
WTA 500	0.031	0.030	0.032	0.023
Ours	**0.013**	**0.009**	**0.010**	**0.008**

Table 5.2.: **Mean Inverse Depth Error.** This table shows the mean inverse depth error for all baseline methods and the proposed method averaged over all 72 test images. The first columns depicts the errors for all pixels where depth ground truth is available, while the other column consider planar regions (of a specific type) only.

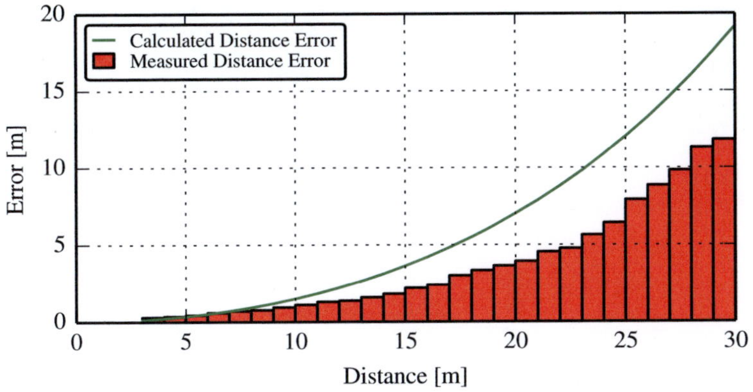

Figure 5.14.: **Reconstruction Error.** This figure shows the exponential dependency between the reconstruction error and the measured distance. The green line denotes the computed mean error from the inverse depth error $0.013\ \mathrm{m}^{-1}$ (see Table 5.2: our method with all pixels) while the red boxes show the mean measured reconstruction error for the different distance ranges.

struction error with the presented method is only 30 cm between 3 m and 4 m and the error for points between 10 m and 11 m is 1.3 m.

5.3.3. Qualitative Results

For the qualitative evaluation we show inverse depth images and the resulting 3D reconstruction with the different analyzed algorithms. In Fig. 5.15 an inverse depth image (top) and the resulting 3D reconstruction (bottom) obtained with the proposed plane-based prior approach on 360° images is shown. Fig. 5.16 and Fig. 5.17 depict a comparison of the results for the different techniques to achieve dense 3D information for two different frames. In (b) - (e) the results for the reference algorithms, Block Matching, Semi-Global Matching, StereoSLIC and winner takes all with threshold 150 are shown. In (f) the results from the proposed approach are presented. Moreover, the ground truth depth maps from the laser scanner (a) (top) and the virtual 360° intensity images (a) (bottom) for the related frame are depicted. In the inverse depth images the color denotes the distance, where green are close and blue distant points. The 3D reconstruction is obtained when reprojecting all pixels of the corresponding inverse depth image back into

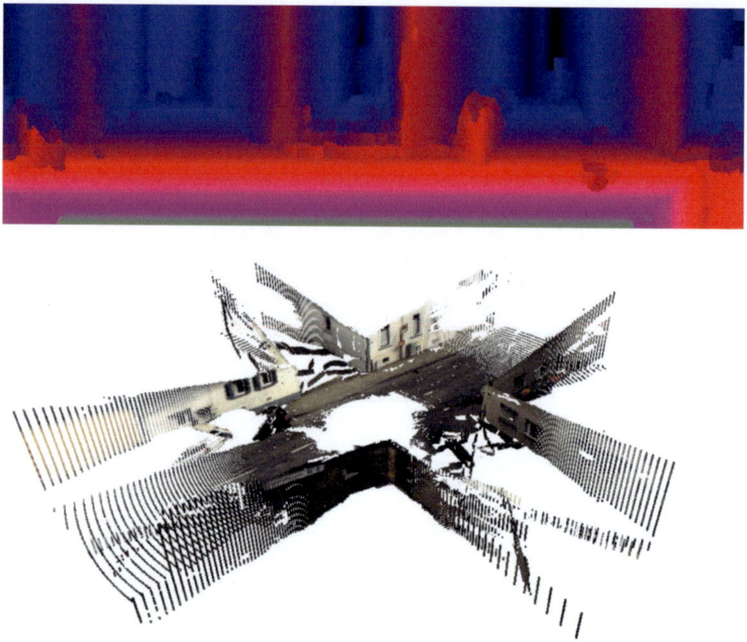

Figure 5.15.: **Dense 3D Reconstruction.** This figure shows the result for the inverse depth image (top) and the resulting 3D reconstruction (bottom) with the proposed plane based model on the virtual 360° image.

3D. A random selection of challenging 3D scenes reconstructed with the proposed method is given in Fig. 5.18.

The proposed approach delivers dense 360° panoramic inverse depth images and a resulting 3D reconstruction of the whole environment. In comparison to the reference methods the depth images are much cleaner and the resulting 3D reconstruction is smoother.

(a) Ground Truth / Velodyne

(b) BM

(c) SGM

(d) StereoSLIC

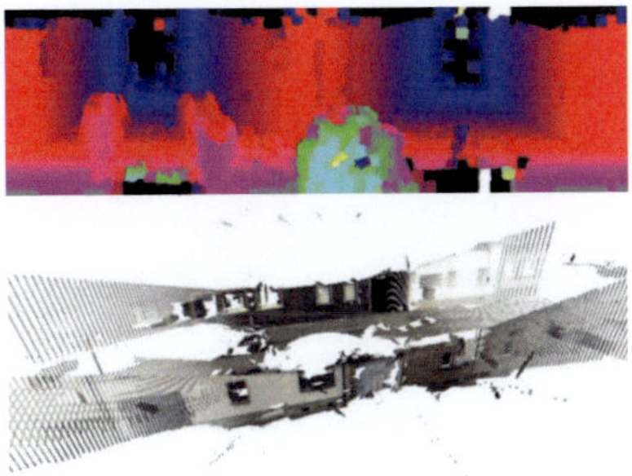

(e) WTA 150

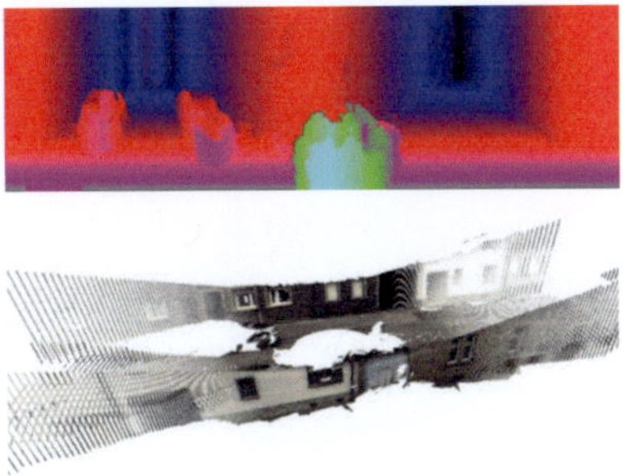

(f) Ours

Figure 5.16.: **Inverse Depth Maps and 3D Reconstructions.** The figure shows the inverse depth images and the resulting 3D reconstruction for the same scene for the baseline algorithms (BM, SGM, StereoSLIC, WTA 150) and the proposed plane based estimation.

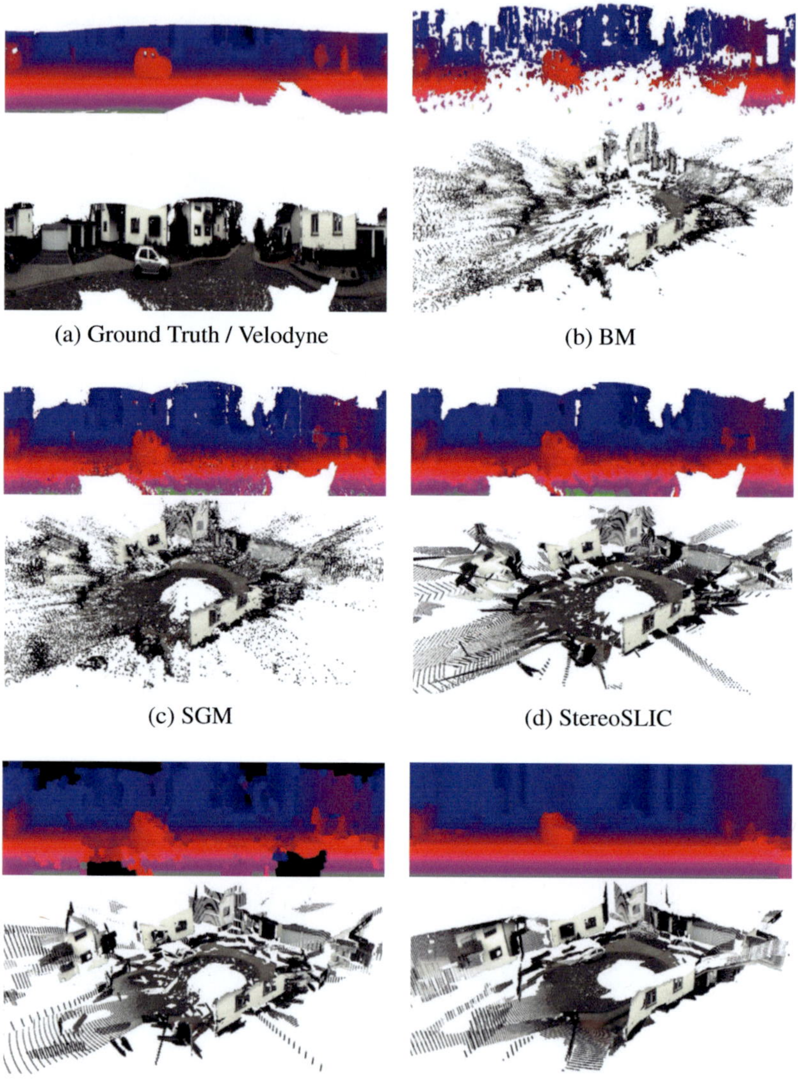

(a) Ground Truth / Velodyne

(b) BM

(c) SGM

(d) StereoSLIC

(e) WTA 150

(f) Ours

Figure 5.17.: **Inverse Depth Maps and 3D Reconstructions.** The figure shows the inverse depth images and the resulting 3D reconstruction for the same scene for the baseline algorithms (BM, SGM, StereoSLIC, WTA 150) and the proposed plane based estimation.

Figure 5.18.: **3D Reconstruction.** This figure shows 3D reconstructions for different urban scenarios obtained when reprojecting the inverse depth maps produced by our proposed plane based method into 3D. Note, that the viewpoint of the rendered 3D point clouds deviates significantly from the viewpoint of the four cameras.

Chapter 6

Conclusion and Outlook

This thesis has proposed a novel stereoscopic omnidirectional camera system for autonomous applications which overcomes the problem of the limited field of view of traditional perspective cameras. Two horizontally aligned catadioptric cameras which provide a $360°$ panoramic view of the environment are used on top of our driving platform. This setup improves many applications for autonomous vehicles which suffer from the limited field of view from perspective cameras. We have shown the potential of our system for two relevant applications, ego-motion estimation and dense 3D reconstruction.

A novel centered projection model for slightly non-central catadioptric cameras has been proposed which is very accurate and has cheep computational costs at the same time. The proposed centered model fills the gap between central models which are efficient but are not accurate enough and non-central models which are accurate but too slow for real-time applications. For the proposed centered projection model once a non-central base model is calibrated to achieve the exact viewing rays. Afterwards, the viewing rays are centered and only the efficient central-centered model is used any time we use a projection function. To compute the parameters of the centered projection model, we have developed a catadioptric stereo calibration toolbox for calibrating multiple catadioptric cameras. This toolbox also allows the calibration of different central reference projection models. We have shown a comparison of the calibration results of the proposed centered projection model against three common central projection models. To demonstrate the advantages of the proposed projection model we have not only analyzed the possibly misleading reprojection error of the calibration targets, but also a localization error of a localization experiment with ground truth has been evaluated. Based on these experiments and the approximation error of the centered projection model, this thesis has shown that the proposed centered projection model approximates non-central catadioptric cameras sufficiently well as long as the distance to the observed points is relatively large compared to the deviation from the single viewpoint which is a reasonable assumption in practice.

As first application an ego-motion estimation method for stereoscopic catadioptric cameras has been presented which overcomes major drawbacks of perspective cameras. We have shown that the motion estimation benefits from the proposed centered projection model and the extended field of view. These benefits have been illustrated with a comparison against the motion estimation with catadioptric cameras using common central projection models and with perspective cameras. Furthermore, a comparative study of feature matching strategies on catadioptric images evaluated against laser-based ground truth has demonstrated that standard feature matching strategies can also be sufficiently used on omnidirectional images. Afterwards, we have presented high fidelity top view maps of the driven path created with the precise ego-motion estimation.

As a second application we show the advantage of the large field of view for dense 3D reconstruction. We have presented a novel method to obtain dense 360° depth images and the resulting 3D reconstruction. The proposed method does not rely on constructing virtual perspective images from the omnidirectional ones and overcomes the depth blind spots by combining motion and spatial stereo. We have shown that planarity priors improve the smoothness of the omnidirectional depth maps and outperform state-of-the art depth estimation techniques in terms of 3D reconstruction error. The 3D reconstruction for different static urban scenes has been exemplary presented.

Regarding further extensions, the proposed setup as well as the efficient and accurate projection model can be used for further applications which also need precise 3D information of the environment such as tracking or localization. For both applications the extended field of view promises an improvement in the results of this tasks. Regarding the dense 3D reconstruction, extensions towards integrating depth information from more than two consecutive stereoscopic views allow for urban reconstruction at larger scales. Moreover, the proposed dense 3D reconstruction combining motion and spatial stereo is not limited to catadioptric images but can be applied on all panoramic images, e.g., panoramic images obtained from multiple fisheye images.

Appendix A

Projection Models

A.1. Geometric Model

Here the computation of the 8th degree polynomial from the two constraints (see Eq. 2.11 and Eq. 2.13) is explained in detail. Note, the points are all represented in the rotated mirror coordinate system. For simplification we omit the index R.

The first constraint can be derived from the law of reflection

$$\mathbf{w}_r = \mathbf{w}_c - \frac{2\mathbf{n}(\mathbf{w}_c^\mathsf{T}\mathbf{n})}{\mathbf{n}^\mathsf{T}\mathbf{n}} \tag{A.1}$$

with

$$\mathbf{n} = \begin{bmatrix} x_m \\ y_m \\ Az_m + B/2 \end{bmatrix}, \quad \mathbf{w}_c = (\mathbf{m} - \mathbf{c}) = \begin{bmatrix} x_m \\ y_m - y_c \\ z_m - z_c \end{bmatrix}$$

and

$$\mathbf{w}_r \times (\mathbf{p} - \mathbf{m}) = \mathbf{w}_r \times \begin{bmatrix} x - x_m \\ y - y_m \\ z - z_m \end{bmatrix} = 0. \tag{A.2}$$

By solving the reflection equation (Eq. A.2), substituting the mirror equation (see Eq. 2.10) $x_m^2 + y_m^2 = C - Az_m^2 - Bz_m$ and examine the first row we achieve

$$I1 : k_{11}(z_m) \cdot y_m^2 + k_{12}(z_m) \cdot y_m + k_{13}(z_m) = 0 \tag{A.3}$$

with

$$\begin{aligned}
k_{11} =& -8y_cAz_m + 8y_cz_m - 8zy_c - 4y_cB \\
k_{12} =& 4yy_cB + 8Az_mC - 4Az_m^2B + 4z_cAz_m^2 - 4z_cA^2z_m^2 + 4zAz_m^2 \\
& - 4zBz_c - 4zA^2z_m^2 + 8yy_cAz_m - 4z_cAz_mB - 8zAz_mz_c - \\
& 4zAz_mB + 4zC + 4z_m^2B - 8z_mC - 2z_mB^2 + 4z_cC - z_cB^2 - \\
& zB^2 + 4BC \\
k_{13} =& -4y_cAz_m^2B - 8yAz_mC + 8yAz_m^2B + 4yz_cAz_m^2 + 4yz_cz_mB + \\
& 4yz_cA^2z_m^2 - 4zy_cAz_m^2 - 4zy_cz_mB + 4zy_cA^2z_m^2 + \\
& 4yz_cAz_mB + 4zy_cAz_mB - 4yAz_m^3 + 4yA^2z_m^3 - 4z_my_cC + \\
& zy_cB^2 - z_my_cB^2 + 4y_cz_m^2B - 4yBC - 4yz_cC - 4y_cA^2z_m^3 + \\
& 3yz_mB^2 + yz_cB^2 - 4yz_m^2B + 4zy_cC + 4yz_mC + 4y_cAz_m^3
\end{aligned}$$

The second constraint is given by

$$(\mathbf{m} - \mathbf{s})^{\mathsf{T}} \cdot \mathbf{n}_{\Pi} = \begin{bmatrix} x_m - 0 \\ y_m - 0 \\ z_m - z_m + Az_m + B/2 \end{bmatrix} \cdot \mathbf{n}_{\Pi} = 0 \tag{A.4}$$

with

$$\mathbf{n}_{\Pi} = (\mathbf{p} - \mathbf{c}) \times (\mathbf{s} - \mathbf{c}) = \begin{pmatrix} x \\ y - y_c \\ z - z_c \end{pmatrix} \times \begin{pmatrix} 0 \\ -y_c \\ -z_c + z_m - Az_m - B/2 \end{pmatrix}$$

From this it follows,

$$c_1(z_m) \cdot x_m + c_2(z_m) \cdot y_m + c_3(z_m) = 0 \tag{A.5}$$

with

$$\begin{aligned}
c_1 &= (B + 2Az_m)(y_c - y) + 2y_c(z - z_m) + 2y(z_m - z_c) \\
c_2 &= x(B + 2z_c - 2z_m + 2Az_m) \\
c_3 &= xy_c(B + 2Az_m)
\end{aligned}$$

Substitute

$$x_m = \frac{-c_2y_m - c_3}{c_1}$$

from Eq. A.5 in the mirror equation $x_m^2 + y_m^2 + Az_m^2 + Bz_m - C = 0$ leads to

$$I2 : \underbrace{(c_1^2 + c_2^2)}_{k_{21}} y_m^2 + \underbrace{2c_2c_3}_{k_{22}} y_m + \underbrace{c_3^2 + c_1^2(Az^2 + Bz - C)}_{k_{23}} = 0 \qquad \text{(A.6)}$$

Combining $I1$ and $I2$ yields

$$\begin{aligned} f(z_m) = \; & k_{21}(z_m)\Big(k_{23}(z_m)k_{12}(z_m)^2 - k_{22}(z_m)k_{12}(z_m)k_{13}(z_m) + \\ & k_{21}(z_m)k_{13}(z_m)^2\Big) - k_{11}(z_m)\Big(- k_{13}(z_m)k_{22}(z_m)^2 + \\ & k_{23}(z_m)k_{12}(z_m)k_{22}(z_m) + 2k_{21}(z_m)k_{23}(z_m)k_{13}(z_m)\Big) + \\ & k_{23}(z_m)^2k_{11}(z_m)^2 = 0. \end{aligned} \qquad \text{(A.7)}$$

Since each k depends quadratic from z_m the polynomial depends on z_m^8.

A.2. Centered Projection Model

Since every central projection model can be represented in the form

$$\begin{bmatrix} u \\ v \end{bmatrix} = \begin{bmatrix} c_u \\ c_v \end{bmatrix} + f_c(\theta) \begin{bmatrix} \cos\varphi \\ \sin\varphi \end{bmatrix}, \qquad \text{(A.8)}$$

where (c_u, c_v) denotes the principal point, φ and θ are the angles of the viewing ray and f_c is an arbitrary monotonic and smooth function, it can be exactly represented by the central-centered projection model. By rearranging the terms

$$\varphi = \arctan\frac{v}{u} \qquad \text{(A.9)}$$

$$\theta = f_c^{-1}\left(\sqrt{(u - c_u)^2 + (v - c_v)^2}\right) \qquad \text{(A.10)}$$

we can compute the new image location with the central-centered projection model from the viewing rays.

Moreover, 3D points at infinity are equivalent mapped to the projection they are derived from. Let $\mathbf{p} = \lambda_c[x, y, z]^\mathsf{T}$ denote a 3D world point and

$\mathbf{t} = [t_x, t_y, t_z]^\mathsf{T}$ an arbitrary finite translation of the viewing ray. Hence, the angles are given as

$$\theta = \arctan \frac{\lambda_c z + t_z}{\sqrt{(\lambda_c x + t_x)^2 + (\lambda_c y + t_y)^2}} \tag{A.11}$$

$$\varphi = \arctan \frac{\lambda_c y + t_y}{\lambda_c x + t_x}. \tag{A.12}$$

For $\lambda_c \to \infty$ we obtain

$$\theta = \arctan \frac{z}{\sqrt{x^2 + y^2}} \tag{A.13}$$

$$\varphi = \arctan \frac{y}{x}. \tag{A.14}$$

Thus, we can represent the viewing ray orientation exactly using the central-centered projection model.

A.3. Perspective Projection Model

For the perspective projection model, we use the normalized projection similar to Eq. 2.18, using the world point $\mathbf{p} = [x, y, z]^\mathsf{T}$ instead of the point on the mirror surface by

$$\mathbf{q}_n^{(P)} = \begin{bmatrix} x_n \\ y_n \end{bmatrix} = \begin{bmatrix} x/z \\ y/z \end{bmatrix}. \tag{A.15}$$

The projected point $\mathbf{q}^{(P)}$ is given by

$$\begin{bmatrix} \mathbf{q}^{(P)} \\ 1 \end{bmatrix} = \underbrace{\begin{bmatrix} f_u & \alpha f_u & c_u \\ 0 & f_v & c_v \\ 0 & 0 & 1 \end{bmatrix}}_{\mathbf{K}} \cdot \begin{bmatrix} \mathbf{q}_d^{(P)} \\ 1 \end{bmatrix} \tag{A.16}$$

where f_u, f_v, c_u, c_v and α are the perspective calibration parameter and $\mathbf{q}_d^{(P)}$ is the distorted point computed from

$$\begin{aligned} \mathbf{q}_d^{(P)} &= \left(1 + k_1 r_n^2 + k_2 r_n^4 + k_5 r_n^6\right) \mathbf{q}_n^{(P)} \\ &+ \begin{bmatrix} 2k_3 x_n y_n + k_4(r_n^2 + 2x_n^2) \\ k_3(r_n^2 + 2y_n^2) + 2k_4 x_n y_n \end{bmatrix} \end{aligned} \tag{A.17}$$

with $r_n = \sqrt{x_n^2 + y_n^2}$ and the distortion parameters $\mathbf{k} = [k_1, \ldots, k_5]^\mathsf{T}$.

Bibliography

[1] G. Adorni, L. Bolognini, S. Cagnoni, and M. Mordonini, "Stereo obstacle detection method for a hybrid omni-directional/pin-hole vision system," in *RoboCup 2001: Robot Soccer World Cup V*. Springer, 2002, pp. 244–250.

[2] A. Agrawal, Y. Taguchi, and S. Ramalingam, "Beyond alhazen's problem: Analytical projection model for non-central catadioptric cameras with quadric mirrors," in *IEEE Conference on Computer Vision and Pattern Recognition (CVPR)*, 2011.

[3] A. Agrawal, Y. Taguchi, and S. Ramalingam, "Analytical forward projection for axial non-central dioptric and catadioptric cameras," in *European Conference on Computer Vision (ECCV)*, 2010.

[4] Z. Arican and P. Frossard, "Scale-invariant features and polar descriptors in omnidirectional imaging," *IEEE Trans. on Image Processing (TIP)*, vol. 21, no. 5, pp. 2412–2423, 2012.

[5] Z. Arican and P. Frossard, "Dense disparity estimation from omnidirectional images." in *IEEE Conference on Advanced Video and Signal Based Surveillance (AVSS)*, 2007.

[6] S. Baker and S. K. Nayar, "A theory of single-viewpoint catadioptric image formation," *International Journal of Computer Vision (IJCV)*, vol. 35, no. 2, pp. 175–196, 1999.

[7] J. P. Barreto and H. Araujo, "Issues on the geometry of central catadioptric image formation," in *IEEE Conference on Computer Vision and Pattern Recognition (CVPR)*, 2001.

[8] J. P. Barreto and H. Araujo, "Geometric properties of central catadioptric line images," in *European Conference on Computer Vision (ECCV)*, 2002.

[9] H. Bay, T. Tuytelaars, and L. V. Gool, "Surf: Speeded up robust features," in *European Conference on Computer Vision (ECCV)*, 2006.

[10] J. Bazin, C. Demonceaux, P. Vasseur, and I. Kweon, "Motion estimation by decoupling rotation and translation in catadioptric vision," *Computer Vision and Image Understanding (CVIU)*, vol. 114, no. 2, pp. 254 – 273, 2010.

[11] R. Benosman and S. B. Kang, *Panoramic Vision : Sensors Theory and Applications*. NY: Springer Verlag, Monographs in Computer Science, ISBN 0-387-95111-3, 2001.

[12] C. M. Bishop, *Pattern Recognition and Machine Learning (Information Science and Statistics)*. Springer-Verlag New York, Inc., 2006.

[13] G. Bradski, *Dr. Dobb's Journal of Software Tools*, 2000.

[14] M. Buehler, K. Iagnemma, and S. Singh, *The DARPA Urban Challenge: Autonomous Vehicles in City Traffic*, 1st ed. Springer, 2009, vol. 56.

[15] R. Bunschoten and B. Kröse, "Visual odometry from an omnidirectional vision system," in *IEEE International Conference on Robotics and Automation (ICRA)*, 2003.

[16] R. Bunschoten, B. J. A. Kröse, and N. A. Vlassis, "Robust scene reconstruction from an omnidirectional vision system." *IEEE Trans. on Robotics and Automation (TRA)*, vol. 19, no. 2, pp. 351–357, 2003.

[17] E. L. Cabral, J. de Souza, and M. C. Hunold, "Omnidirectional stereo vision with a hyperbolic double lobed mirror," in *International Conference on Pattern Recognition (ICPR)*, 2004.

[18] V. Caglioti, P. Taddei, G. Boracchi, S. Gasparini, and A. Giusti, "Single-image calibration of off-axis catadioptric cameras using lines," in *Workshop on Omnidirectional Vision, Camera Networks and Nonclassical cameras (OMNIVIS)*, 2007.

[19] M. Calonder, V. Lepetit, M. Ozuysal, T. Trzcinski, C. Strecha, and P. Fua, "Brief: Computing a local binary descriptor very fast," *IEEE Trans. on Pattern Analysis and Machine Intelligence (PAMI)*, vol. 34, no. 7, pp. 1281–1298, 2012.

[20] J. S. Chahl and M. V. Srinivasan, "Reflective surfaces for panoramic imaging," *Applied Optics*, vol. 36, no. 31, pp. 8275–8285, 1997.

[21] P. Chang and M. Hebert, "Omni-directional structure from motion," in *Workshop on Omnidirectional Vision, Camera Networks and Non-classical cameras (OMNIVIS)*, 2000.

[22] S. Y. Cheng and M. M. Trivedi, "Toward a comparative study of lane tracking using omni-directional and rectilinear images for driver assistance systems," in *IEEE International Conference on Robotics and Automation (ICRA)*, 2007.

[23] P. Corke, D. Strelow, and S. Singh, "Omnidirectional visual odometry for a planetary rover," in *IEEE International Conference on Intelligent Robots and Systems (IROS)*, 2004.

[24] M. Cummins and P. Newman, "Probabilistic appearance based navigation and loop closing," in *IEEE International Conference on Robotics and Automation (ICRA)*, 2007.

[25] R. Descartes, *The Geometry of René Descartes*. Courier Dover Publications, 2012.

[26] E. Dickmanns, R. Behringer, D. Dickmanns, T. Hildebrandt, M. Maurer, F. Thomanek, and J. Schiehlen, "The seeing passenger car 'vamors-p'," in *IEEE Intelligent Vehicles Symposium (IV)*, 1994.

[27] H. Durrant-Whyte and T. Bailey, "Simultaneous localization and mapping: part i," *IEEE Robotics and Automation Magazine*, vol. 13, no. 2, pp. 99–110, June 2006.

[28] T. Ehlgen, T. Pajdla, and D. Ammon, "Eliminating blind spots for assisted driving," *IEEE Trans. on Intelligent Transportation Systems (TITS)*, vol. 9, no. 4, pp. 657–665, 2008.

[29] T. Ehlgen, M. Thom, and M. Glaser, "Omnidirectional cameras as backing-up aid," in *IEEE International Conference on Computer Vision (ICCV)*, 2007.

[30] M. Fiala and A. Basu, "Feature extraction and calibration for stereo reconstruction using non-svp optics in a panoramic stereo-vision sensor," in *Workshop on Omnidirectional Vision, Camera Networks and Nonclassical cameras (OMNIVIS)*, 2002.

[31] M. Fiala and A. Basu, "Panoramic stereo reconstruction using non-svp optics," *Computer Vision and Image Understanding (CVIU)*, vol. 98, no. 3, pp. 363–397, 2005.

[32] M. A. Fischler and R. C. Bolles, "Random sample consensus: a paradigm for model fitting with applications to image analysis and automated cartography," *Communications of the ACM*, vol. 24, no. 6, pp. 381–395, 1981.

[33] S. Fleck, F. Busch, P. Biber, W. Strasser, and H. Andreasson, "Omnidirectional 3d modeling on a mobile robot using graph cuts," in *IEEE International Conference on Robotics and Automation (ICRA)*, 2005.

[34] J. Fujiki, A. Torii, and S. Akaho, "Epipolar geometry via rectification of spherical images," in *Computer Vision/Computer Graphics Collaboration Techniques*. Springer, 2007, pp. 461–471.

[35] Y. Furukawa, B. Curless, S. M. Seitz, and R. Szeliski, "Manhattan-world stereo," in *IEEE Conference on Computer Vision and Pattern Recognition (CVPR)*, 2009.

[36] Y. Furukawa, B. Curless, S. M. Seitz, and R. Szeliski, "Reconstructing building interiors from images," in *IEEE International Conference on Computer Vision (ICCV)*, 2009.

[37] T. Gandhi and M. Trivedi, "Vehicle surround capture: Survey of techniques and a novel omni-video-based approach for dynamic panoramic surround maps," *IEEE Trans. on Intelligent Transportation Systems (TITS)*, vol. 7, no. 3, pp. 293–308, 2006.

[38] T. Gandhi and M. Trivedi, "Video based surround vehicle detection, classification and logging from moving platforms: Issues and approaches," in *IEEE Intelligent Vehicles Symposium (IV)*, 2007.

[39] J. Gaspar, C. Deccó, J. Okamoto, and J. Santos-Victor, "Constant resolution omnidirectional cameras," in *Workshop on Omnidirectional Vision, Camera Networks and Nonclassical cameras (OMNIVIS)*, 2002.

[40] S. K. Gehrig, "Large-field-of-view stereo for automotive applications," *Workshop on Omnidirectional Vision, Camera Networks and Nonclassical cameras (OMNIVIS)*, 2005.

[41] A. Geiger, P. Lenz, and R. Urtasun, "Are we ready for autonomous driving? the kitti vision benchmark suite," in *IEEE Conference on Computer Vision and Pattern Recognition (CVPR)*, 2012.

[42] A. Geiger, F. Moosmann, O. Car, and B. Schuster, "A toolbox for automatic calibration of range and camera sensors using a single shot," in *IEEE International Conference on Robotics and Automation (ICRA)*, 2012.

[43] A. Geiger, J. Ziegler, and C. Stiller, "StereoScan: Dense 3d reconstruction in real-time," in *IEEE Intelligent Vehicles Symposium (IV)*, 2011.

[44] C. Geyer and K. Daniilidis, "Paracatadioptric camera calibration," *IEEE Trans. on Pattern Analysis and Machine Intelligence (PAMI)*, vol. 24, no. 5, pp. 687–695, 2002.

[45] C. Geyer and K. Daniilidis, "A unifying theory for central panoramic systems and practical implications," in *European Conference on Computer Vision (ECCV)*, 2000.

[46] C. Geyer and K. Daniilidis, "Conformal rectification of omnidirectional stereo pairs," in *Workshop on Omnidirectional Vision, Camera Networks and Nonclassical cameras (OMNIVIS)*, 2003.

[47] J. Gluckman and S. K. Nayar, "Ego-motion and omnidirectional cameras," in *IEEE International Conference on Computer Vision (ICCV)*, 1998.

[48] J. Gluckman, S. K. Nayar, and K. J. Thoresz, "Real-time omnidirectional and panoramic stereo," in *In DARPA Image Understanding Workshop*, 1998.

[49] N. Gonçalves and A. Nogueira, "Projection through quadric mirrors made faster," in *Workshop on Omnidirectional Vision, Camera Networks and Nonclassical cameras (OMNIVIS)*, 2009.

[50] J.-J. Gonzalez-Barbosa and S. Lacroix, "Fast dense panoramic stereovision." in *IEEE International Conference on Robotics and Automation (ICRA)*, 2005.

[51] M. D. Grossberg and S. K. Nayar, "A general imaging model and a method for finding its parameters," in *IEEE International Conference on Computer Vision (ICCV)*, 2001.

[52] P. Hansen, P. Corke, W. Boles, and K. Daniilidis, "Scale invariant feature matching with wide angle images," in *IEEE International Conference on Intelligent Robots and Systems (IROS)*, 2007.

[53] R. I. Hartley and A. Zisserman, *Multiple View Geometry in Computer Vision*, 2nd ed. Cambridge University Press, 2004.

[54] L. He, C. Luo, F. Zhu, Y. Hao, J. Ou, and J. Zhou, "Depth map regeneration via improved graph cuts using a novel omnidirectional stereo sensor," in *IEEE International Conference on Computer Vision (ICCV)*, 2007.

[55] R. A. Hicks and R. Bajcsy, "Catadioptric sensors that approximate wide-angle perspective projections." in *IEEE Conference on Computer Vision and Pattern Recognition (CVPR)*, 2000.

[56] H. Hirschmüller, "Stereo processing by semiglobal matching and mutual information," *IEEE Trans. on Pattern Analysis and Machine Intelligence (PAMI)*, vol. 30, no. 2, pp. 328–341, 2008.

[57] H.-C. Huang and Y.-P. Hung, "Panoramic stereo imaging system with automatic disparity warping and seaming," *Graphical Models and Image Processing*, vol. 60, no. 3, pp. 196–208, 1998.

[58] K. S. Huang, M. M. Trivedi, and T. Gandhi, "Driver's view and vehicle surround estimation using omnidirectional video stream," in *IEEE Intelligent Vehicles Symposium (IV)*, 2003.

[59] H. Ishiguro, M. Yamamoto, and S. Tsuji, "Omni-directional stereo," *IEEE Trans. on Pattern Analysis and Machine Intelligence (PAMI)*, vol. 14, no. 2, pp. 257–262, Feb 1992.

[60] G. Jang, S. Kim, and I. Kweon, "Single camera catadioptric stereo system," in *Workshop on Omnidirectional Vision, Camera Networks and Nonclassical cameras (OMNIVIS)*, 2005.

[61] N. Jankovic and M. Naish, "A centralized omnidirectional multi-camera system with peripherally-guided active vision and depth perception," in *IEEE International Conference on Networking, Sensing and Control*, 2007.

[62] H. L. Jasper Snoek and R. P. Adams, "Practical bayesian optimization of machine learning algorithms," in *Advances in Neural Information Processing Systems (NIPS)*, 2012.

[63] S. B. Kang, "Catadioptric self-calibration," in *IEEE Conference on Computer Vision and Pattern Recognition (CVPR)*, 2000.

[64] S. B. Kang and R. Szeliski, "3-d scene data recovery using omnidirectional multibaseline stereo," *International Journal of Computer Vision (IJCV)*, vol. 25, no. 2, pp. 167–183, 1995.

[65] M. Kobilarov, G. Sukhatme, J. Hyams, and P. Batavia, "People tracking and following with mobile robot using an omnidirectional camera and a laser," in *IEEE International Conference on Robotics and Automation (ICRA)*, 2006.

[66] H. Lategahn and C. Stiller, "Vision-only localization," *IEEE Trans. on Intelligent Transportation Systems (TITS)*, vol. 15, no. 3, pp. 1246–1257, June 2014.

[67] M. Lauer, M. Schönbein, S. Lange, and S. Welker, "3d-objecttracking with a mixed omnidirectional stereo camera system," *Mechatronics*, vol. 21, no. 2, pp. 390 – 398, 2011.

[68] S. Leutenegger, M. Chli, and R. Y. Siegwart, "Brisk: Binary robust invariant scalable keypoints," in *IEEE International Conference on Computer Vision (ICCV)*, 2011.

[69] M. Lhuillier, "Toward flexible 3d modeling using a catadioptric camera," in *IEEE Conference on Computer Vision and Pattern Recognition (CVPR)*, 2007.

[70] M. Lhuillier, "Automatic scene structure and camera motion using a catadioptric system," *Computer Vision and Image Understanding (CVIU)*, vol. 109, no. 2, pp. 186–203, 2008.

[71] S. Li, "Real-time spherical stereo," in *International Conference on Pattern Recognition (ICPR)*, 2006.

[72] D. G. Lowe, "Distinctive image features from scale-invariant keypoints," *International Journal of Computer Vision (IJCV)*, vol. 60, no. 2, pp. 91–110, 2004.

[73] D. W. Marquardt, "An algorithm for least-squares estimation of nonlinear parameters," *Journal of the Society for Industrial & Applied Mathematics*, vol. 11, no. 2, pp. 431–441, 1963.

[74] L. Matuszyk, A. Zelinsky, L. Nilsson, and M. Rilbe, "Stereo panoramic vision for monitoring vehicle blind-spots," in *IEEE Intelligent Vehicles Symposium (IV)*, 2004.

[75] C. Mei and P. Rives, "Single view point omnidirectional camera calibration from planar grids," in *IEEE International Conference on Robotics and Automation (ICRA)*, 2007.

[76] B. Mičušík and J. Košecká, "Multi-view superpixel stereo in urban environments," *International Journal of Computer Vision (IJCV)*, vol. 89, no. 1, pp. 106–119, 2010.

[77] B. Mičušík and T. Pajdla, "Autocalibration & 3d reconstruction with non-central catadioptric cameras," in *IEEE Conference on Computer Vision and Pattern Recognition (CVPR)*, 2004.

[78] B. Mičušík and T. Pajdla, "Structure from motion with wide circular field of view cameras," *IEEE Trans. on Pattern Analysis and Machine Intelligence (PAMI)*, vol. 28, no. 7, pp. 1135–1149, 2006.

[79] K. Mikolajczyk and C. Schmid, "A performance evaluation of local descriptors," *IEEE Trans. on Pattern Analysis and Machine Intelligence (PAMI)*, vol. 27, no. 10, pp. 1615–1630, 2005.

[80] O. Miksik and K. Mikolajczyk, "Evaluation of local detectors and descriptors for fast feature matching," in *International Conference on Pattern Recognition (ICPR)*, 2012.

[81] O. Morel, R. Seulin, and D. Fofi, "Catadioptric camera calibration by polarization imaging," in *Pattern Recognition and Image Analysis*. Springer, 2007, pp. 396–403.

[82] A. Murillo, J. Guerrero, and C. Sagues, "Surf features for efficient robot localization with omnidirectional images," in *IEEE International Conference on Robotics and Automation (ICRA)*, 2007.

[83] S. Nayar, "Catadioptric omnidirectional camera," in *IEEE Conference on Computer Vision and Pattern Recognition (CVPR)*, 1997.

[84] A. Neubeck and L. V. Gool, "Efficient non-maximum suppression," in *International Conference on Pattern Recognition (ICPR)*, 2006.

[85] K. Ng, H. Ishiguro, M. Trivedi, and T. Sogo, "Monitoring dynamically changing environments by ubiquitous vision system," in *Second IEEE Workshop on Visual Surveillance (VS)*, 1999.

[86] D. Nister, O. Naroditsky, and J. Bergen, "Visual odometry," in *IEEE Conference on Computer Vision and Pattern Recognition (CVPR)*, 2004.

[87] S. Peleg, M. Ben-Ezra, and Y. Pritch, "Omnistereo: panoramic stereo imaging," *IEEE Trans. on Pattern Analysis and Machine Intelligence (PAMI)*, vol. 23, no. 3, pp. 279–290, 2001.

[88] L. Puig, Y. Bastanlar, P. Sturm, J. Guerrero, and J. Barreto, "Calibration of central catadioptric cameras using a dlt-like approach," *International Journal of Computer Vision (IJCV)*, vol. 93, pp. 101–114, 2011.

[89] L. Puig, J. Bermúdez, P. Sturm, and J. J. Guerrero, "Calibration of omnidirectional cameras in practice: A comparison of methods," *Computer Vision and Image Understanding (CVIU)*, 2012.

[90] S. Ramalingam, P. Sturm, and S. K. Lodha, "Towards complete generic camera calibration," in *IEEE Conference on Computer Vision and Pattern Recognition (CVPR)*, 2005.

[91] D. W. Rees, "Panoramic television viewing system," Apr. 7 1970, uS Patent 3,505,465.

[92] A. Rituerto, L. Puig, and J. Guerrero, "Comparison of omnidirectional and conventional monocular systems for visual slam," *Workshop on Omnidirectional Vision, Camera Networks and Nonclassical cameras (OMNIVIS)*, 2010.

[93] E. Rosten and T. Drummond, "Machine learning for high-speed corner detection," in *European Conference on Computer Vision (ECCV)*, 2006.

[94] E. Rublee, V. Rabaud, K. Konolige, and G. Bradski, "Orb: an efficient alternative to sift or surf," in *IEEE International Conference on Computer Vision (ICCV)*, 2011.

[95] D. Scaramuzza, F. Fraundorfer, and R. Siegwart, "Real-time monocular visual odometry for on-road vehicles with 1-point ransac," in *IEEE International Conference on Robotics and Automation (ICRA)*, 2009.

[96] D. Scaramuzza, A. Martinelli, and R. Siegwart, "A toolbox for easy calibrating omnidirectional cameras," in *IEEE International Conference on Intelligent Robots and Systems (IROS)*, 2006.

[97] D. Scaramuzza, A. Harati, and R. Siegwart, "Extrinsic self calibration of a camera and a 3d laser range finder from natural scenes," in *IEEE International Conference on Intelligent Robots and Systems (IROS)*. IEEE, 2007, pp. 4164–4169.

[98] D. Scaramuzza, A. Martinelli, and R. Siegwart, "A flexible technique for accurate omnidirectional camera calibration and structure from motion," in *IEEE International Conference on Computer Vision (ICCV)*, 2006.

[99] D. Scaramuzza, A. Martinelli, and Y. Siegwart, Roland, "A robust descriptor for tracking vertical lines in omnidirectional images and its use in mobile robotics," *International Journal of Robotics Research (IJRR)*, vol. 28, no. 2, pp. 149–171, 2009.

[100] D. Scaramuzza and R. Siegwart, "Appearance-guided monocular omnidirectional visual odometry for outdoor ground vehicles," *IEEE Trans. on Robotics*, vol. 24, no. 5, pp. 1015–1026, 2008.

[101] D. Scharstein and R. Szeliski, "A taxonomy and evaluation of dense two-frame stereo correspondence algorithms," *International Journal of Computer Vision (IJCV)*, vol. 47, no. 1-3, pp. 7–42, 2002.

[102] G. Schindler and F. Dellaert, "Atlanta world: An expectation maximization framework for simultaneous low-level edge grouping and camera calibration in complex man-made environments," in *IEEE Conference on Computer Vision and Pattern Recognition (CVPR)*, 2004.

[103] M. Schönbein and A. Geiger, "Omnidirectional 3d reconstruction in augmented manhattan worlds," in *IEEE International Conference on Intelligent Robots and Systems (IROS)*, 2014.

[104] M. Schönbein, B. Kitt, and M. Lauer, "Environmental perception for intelligent vehicles using catadioptric stereo vision systems," in *Proc. of the European Conference on Mobile Robots (ECMR)*, 2011.

[105] M. Schönbein, H. Rapp, and M. Lauer, "Panoramic 3d reconstruction with three catadioptric cameras," in *Intelligent Autonomous Systems 12*. Springer, 2013, pp. 345–353.

[106] M. Schönbein, T. Strauss, and A. Geiger, "Calibrating and centering quasi-central catadioptric cameras," in *IEEE International Conference on Robotics and Automation (ICRA)*, 2014.

[107] A. Schwing and R. Urtasun, "Efficient exact inference for 3d indoor scene understanding," in *European Conference on Computer Vision (ECCV)*, 2012.

[108] T. Sogo, H. Ishiguro, and M. M. Trivedi, “Real-time target localization and tracking by n-ocular stereo,” in *Workshop on Omnidirectional Vision, Camera Networks and Nonclassical cameras (OMNIVIS)*, 2000.

[109] D. Strelow, J. S. Mishler, D. Koes, and S. Singh, “Precise omnidirectional camera calibration,” in *IEEE Conference on Computer Vision and Pattern Recognition (CVPR)*, 2001.

[110] P. Sturm, “Mixing catadioptric and perspective cameras,” in *Workshop on Omnidirectional Vision, Camera Networks and Nonclassical cameras (OMNIVIS)*, 2002.

[111] P. Sturm and S. Ramalingam, “A generic concept for camera calibration,” in *European Conference on Computer Vision (ECCV)*, 2004.

[112] P. Sturm, S. Ramalingam, J.-P. Tardif, S. Gasparini, and J. a. Barreto, “Camera models and fundamental concepts used in geometric computer vision,” *Foundations and Trends® in Computer Graphics and Vision*, vol. 6, no. 1–2, pp. 1–183, 2011.

[113] T. Svoboda and T. Pajdla, “Epipolar geometry for central catadioptric cameras,” *International Journal of Computer Vision (IJCV)*, vol. 49, no. 1, pp. 23–37, 2002.

[114] R. Swaminathan, M. Grossberg, and S. Nayar, “Caustics of catadioptric cameras,” in *IEEE International Conference on Computer Vision (ICCV)*, 2001.

[115] R. Swaminathan, M. D. Grossberg, and S. K. Nayar, “Non-single viewpoint catadioptric cameras: Geometry and analysis,” *International Journal of Computer Vision (IJCV)*, vol. 66, no. 3, pp. 211–229, 2006.

[116] J. Takiguchi, M. Yoshida, A. Takeya, J. Eino, and T. Hashizume, “High precision range estimation from an omnidirectional stereo system,” in *IEEE/RSJ International Conference on Intelligent Robots and Systems*, 2002.

[117] J.-P. Tardif, Y. Pavlidis, and K. Daniilidis, “Monocular visual odometry in urban environments using an omnidirectional camera,” in *IEEE International Conference on Intelligent Robots and Systems (IROS)*, 2008.

[118] B. Triggs, P. F. McLauchlan, R. I. Hartley, and A. W. Fitzgibbon, "Bundle adjustment – a modern synthesis," in *Vision algorithms: theory and practice*. Springer, 2000, pp. 298–372.

[119] M. Trivedi, T. Gandhi, and J. McCall, "Looking-in and looking-out of a vehicle: Computer-vision-based enhanced vehicle safety," *IEEE Trans. on Intelligent Transportation Systems (TITS)*, vol. 8, no. 1, pp. 108–120, 2007.

[120] C. Valgren and A. J. Lilienthal, "Sift, surf & seasons: Appearance-based long-term localization in outdoor environments," *Robotics and Autonomous Systems (RAS)*, vol. 58, no. 2, pp. 149–156, 2010.

[121] A. Voigtländer, S. Lange, M. Lauer, and M. A. Riedmiller, "Real-time 3d ball recognition using perspective and catadioptric cameras." in *Proc. European Conference on Mobile Robotics (ECMR)*, 2007.

[122] H. Winner, S. Hakuli, and G. Wolf, *Handbuch Fahrerassistenzsysteme: Grundlagen, Komponenten und Systeme für aktive Sicherheit und Komfort*. Vieweg+Teubner, 2009.

[123] Y. Yagi, S. Kawato, and S. Tsuji, "Real-time omnidirectional image sensor (copis) for vision-guided navigation," *IEEE Trans. on Robotics and Automation (TRA)*, vol. 10, no. 1, pp. 11–22, 1994.

[124] K. Yamaguchi, D. McAllester, and R. Urtasun, "Robust monocular epipolar flow estimation," *IEEE Conference on Computer Vision and Pattern Recognition (CVPR)*, 2013.

[125] S. Yi and N. Ahuja, "An omnidirectional stereo vision system using a single camera," in *International Conference on Pattern Recognition (ICPR)*, 2006.

[126] X. Ying and Z. Hu, "Can we consider central catadioptric cameras and fisheye cameras within a unified imaging model," in *European Conference on Computer Vision (ECCV)*, 2004.

[127] X. Ying and Z. Hu, "Catadioptric camera calibration using geometric invariants," *IEEE Trans. on Pattern Analysis and Machine Intelligence (PAMI)*, vol. 26, pp. 1260–1271, 2004.

[128] S. Yu and M. Lhuillier, "Surface reconstruction of scenes using a catadioptric camera," in *Computer Vision/Computer Graphics Collaboration Techniques*. Springer, 2011, pp. 145–156.

[129] B. Zeisl, C. Zach, and M. Pollefeys, "Stereo reconstruction of building interiors with a vertical structure prior," in *Proc. of the International Conf. on 3D Digital Imaging, Modeling, Data Processing, Visualization and Transmission (THREEDIMPVT)*, 2011.

[130] Z. Zhu, "Omnidirectional stereo vision," in *Workshop on Omnidirectional Vision, Camera Networks and Nonclassical cameras (OMNIVIS)*, 2001.

[131] Z. Zhu, K. D. Rajasekar, E. M. Riseman, and A. R. Hanson, "Panoramic virtual stereo vision of cooperative mobile robots for localizing 3d moving objects," in *Workshop on Omnidirectional Vision, Camera Networks and Nonclassical cameras (OMNIVIS)*, 2000.

[132] J. Ziegler, P. Bender, M. Schreiber, H. Lategahn, T. Strauss, C. Stiller, T. Dang, U. Franke, N. Appenrodt, C. Keller, E. Kaus, R. Herrtwich, C. Rabe, D. Pfeiffer, F. Lindner, F. Stein, F. Erbs, M. Enzweiler, C. Knöppel, J. Hipp, M. Haueis, M. Trepte, C. Brenk, A. Tamke, M. Ghanaat, M. Braun, A. Joos, H. Fritz, H. Mock, M. Hein, and E. Zeeb, "Making bertha drive - an autonomous journey on a historic route," *IEEE Intelligent Transportation Systems Magazine*, vol. 6, no. 2, pp. 8–20, 2014.

Schriftenreihe
Institut für Mess- und Regelungstechnik
Karlsruher Institut für Technologie
(1613-4214)

Die Bände sind unter www.ksp.kit.edu als PDF frei verfügbar
oder als Druckausgabe bestellbar.

Band 001 Hans, Annegret
Entwicklung eines Inline-Viskosimeters auf Basis eines magnetisch-induktiven Durchflussmessers. 2004
ISBN 3-937300-02-3

Band 002 Heizmann, Michael
Auswertung von forensischen Riefenspuren mittels automatischer Sichtprüfung. 2004
ISBN 3-937300-05-8

Band 003 Herbst, Jürgen
Zerstörungsfreie Prüfung von Abwasserkanälen mit Klopfschall. 2004
ISBN 3-937300-23-6

Band 004 Kammel, Sören
Deflektometrische Untersuchung spiegelnd reflektierender Freiformflächen. 2005
ISBN 3-937300-28-7

Band 005 Geistler, Alexander
Bordautonome Ortung von Schienenfahrzeugen mit Wirbelstrom-Sensoren. 2007
ISBN 978-3-86644-123-1

Band 006 Horn, Jan
Zweidimensionale Geschwindigkeitsmessung texturierter Oberflächen mit flächenhaften bildgebenden Sensoren. 2007
ISBN 978-3-86644-076-0

Band 007 Hoffmann, Christian
Fahrzeugdetektion durch Fusion monoskopischer Videomerkmale. 2007
ISBN 978-3-86644-139-2

Band 008 Dang, Thao
Kontinuierliche Selbstkalibrierung von Stereokameras. 2007
ISBN 978-3-86644-164-4

Band 009 Kapp, Andreas
Ein Beitrag zur Verbesserung und Erweiterung der Lidar-Signalverarbeitung für Fahrzeuge. 2007
ISBN 978-3-86644-174-3

Band 010 Horbach, Jan
Verfahren zur optischen 3D-Vermessung spiegelnder Oberflächen. 2008
ISBN 978-3-86644-202-3

Band 011 Böhringer, Frank
Gleisselektive Ortung von Schienenfahrzeugen mit bordautonomer Sensorik. 2008
ISBN 978-3-86644-196-5

Band 012 Xin, Binjian
Auswertung und Charakterisierung dreidimensionaler Messdaten technischer Oberflächen mit Riefentexturen. 2009
ISBN 978-3-86644-326-6

Band 013 Cech, Markus
Fahrspurschätzung aus monokularen Bildfolgen für innerstädtische Fahrerassistenzanwendungen. 2009
ISBN 978-3-86644-351-8

Band 014 Speck, Christoph
Automatisierte Auswertung forensischer Spuren auf Patronenhülsen. 2009
ISBN 978-3-86644-365-5

Band 015 Bachmann, Alexander
Dichte Objektsegmentierung in Stereobildfolgen. 2010
ISBN 978-3-86644-541-3

Band 016 Duchow, Christian
Videobasierte Wahrnehmung markierter Kreuzungen mit lokalem Markierungstest und Bayes'scher Modellierung. 2011
ISBN 978-3-86644-630-4

Band 017 Pink, Oliver
Bildbasierte Selbstlokalisierung von Straßenfahrzeugen. 2011
ISBN 978-3-86644-708-0

Band 018 Hensel, Stefan
Wirbelstromsensorbasierte Lokalisierung von Schienenfahrzeugen in topologischen Karten. 2011
ISBN 978-3-86644-749-3

Band 019 Carsten Hasberg
Simultane Lokalisierung und Kartierung spurgeführter Systeme. 2012
ISBN 978-3-86644-831-5

Band 020 Pitzer, Benjamin
Automatic Reconstruction of Textured 3D Models. 2012
ISBN 978-3-86644-805-6

Band 021 Roser, Martin
Modellbasierte und positionsgenaue Erkennung von Regentropfen in Bildfolgen zur Verbesserung von videobasierten Fahrerassistenzfunktionen. 2012
ISBN 978-3-86644-926-8

Band 022 Loose, Heidi
Dreidimensionale Straßenmodelle für Fahrerassistenzsysteme auf Landstraßen. 2013
ISBN 978-3-86644-942-8

Band 023 Rapp, Holger
Reconstruction of Specular Reflective Surfaces using Auto-Calibrating Deflectometry. 2013
ISBN 978-3-86644-966-4

Band 024 Moosmann, Frank
Interlacing Self-Localization, Moving Object Tracking and Mapping for 3D Range Sensors. 2013
ISBN 978-3-86644-977-0

Band 025 Geiger, Andreas
Probabilistic Models for 3D Urban Scene Understanding from Movable Platforms. 2013
ISBN 978-3-7315-0081-0

Band 026 Hörter, Marko
Entwicklung und vergleichende Bewertung einer bildbasierten Markierungslichtsteuerung für Kraftfahrzeuge. 2013
ISBN 978-3-7315-0091-9

Band 027 Kitt, Bernd
Effiziente Schätzung dichter Bewegungsvektorfelder unter Berücksichtigung der Epipolargeometrie zwischen unterschiedlichen Ansichten einer Szene. 2013
ISBN 978-3-7315-0105-3

Band 028 Lategahn, Henning
Mapping and Localization in Urban Environments Using Cameras. 2013
ISBN 978-3-7315-0135-0

Band 029 Tischler, Karin
Informationsfusion für die kooperative Umfeldwahrnehmung vernetzter Fahrzeuge. 2014
ISBN 978-3-7315-0166-4

Band 030 Schmidt, Christian
Fahrstrategien zur Unfallvermeidung im Straßenverkehr für Einzel- und Mehrobjektszenarien. 2014
ISBN 978-3-7315-0198-5

Band 031 Firl, Jonas
Probabilistic Maneuver Recognition in Traffic Scenarios. 2014
ISBN 978-3-7315-0287-6

Band 032 Schönbein, Miriam
Omnidirectional Stereo Vision for Autonomous Vehicles. 2015
ISBN 978-3-7315-0357-6